MODERN ELEMENTARY DIFFERENTIAL EQUATIONS

RICHARD BELLMAN
Late Professor of Mathematics,
University of Southern California

and

KENNETH L. COOKE
Emeritus Professor of Mathematics,
Pomona College, Claremont, California

SECOND EDITION

DOVER PUBLICATIONS, INC.
New York

Published in Canada by General Publishing Company, Ltd., 30 Lesmill Road, Don Mills, Toronto, Ontario.

Bibliographical Note

This Dover edition, first published in 1995, is an unabridged, slightly corrected republication of the second edition (1971, revised) of the work originally published by Addison-Wesley Publishing Company, Reading, Mass., 1968.

Library of Congress Cataloging-in-Publication Data

Bellman, Richard Ernest, 1920–
Modern elementary differential equations / Richard Bellman and Kenneth L. Cooke. — 2nd ed.
p. cm.
"An unabridged, slightly corrected republication of the second edition (1971, revised) of the work originally published by Addison-Wesley Publishing, Reading, 1968"—T.p. verso.
Includes bibliographical references (p. –) and index.
ISBN 0-486-68643-4 (pbk.)
1. Differential equations. I. Cooke, Kenneth L. II. Title.
QA372.B418 1995
515′.35—dc20 94-49360
CIP

Manufactured in the United States of America
Dover Publications, Inc., 31 East 2nd Street, Mineola, N.Y. 11501

Preface to the Second Edition

Our objective in preparing this revised edition has been to enhance the usefulness of this text to the student. We have attempted to make the book more easily readable by clarifying terminology and rearranging and adding material. New sections have been included on a number of important topics, some of which were previously contained only in the problem lists. For example, Chapter 2 contains new sections on discontinuous solutions, the Riccati and Euler equations, and linear difference equations. Alternative techniques for generating series solutions have been added in Chapter 3, and a discussion of the function evaluations required in the numerical solution of differential equations has been added to Chapter 4. As a further aid to the student, many new exercises and problems are included and a list of answers to selected problems has been provided. We wish to thank Mr. Art Lew for his assistance in preparing the answers.

Los Angeles and Claremont, California
October, 1970

R.B.
K.L.C.

Preface to the First Edition

The theory of differential equations plays a fundamental role in both pure and applied mathematics. Indeed, some of the most important ideas in analysis, algebra, and topology were developed in attempts of mathematicians to resolve particular problems involving equations of this nature. The contemporary efforts to introduce quantitative and qualitative measures into such intellectual disciplines as biology, economics, psychology, and other social sciences has considerably enhanced this role of differential equations and will undoubtedly further enrich mathematics. The startling growth of electronic computers which can solve large systems of differential equations quickly and accurately has made the formulation of scientific problems in terms of differential equations more meaningful than ever before.

It is therefore essential that a thorough grounding in the theory and application of differential equations be a part of the scientific education of every scientist, both physical and social. Toward this end, it will be necessary to "retread" a number of practicing engineers and scientists who were educated before the day of the routine use of computers. This book is directed toward both these objectives, and is intended for use in an introductory course. We assume only an elementary knowledge of calculus, including the rudiments of the theory of power series.

The question of what material to include in a text of this nature and how to present it is not an easy one. In organizing the material, we have two objectives in mind. To begin with, we want the reader to possess certain knowledge and skills which could be used in companion courses in engineering, physics, biology, economics, and so on. Secondly, we want him to be aware of the many mathematical problems which arise in a course in applications and thus be motivated to take many further courses in the theory of differential equations, or to pursue these studies on his own.

We feel that the following are minimum objectives for the basic training of a student:

1. An understanding of how to use differential equations in order to describe a variety of physical processes. By this we mean that the reader should understand how to convert plausible scientific assumptions into various kinds of differential equations. A part of this understanding is an appreciation of when differential equations alone do not suffice.

2. An ability to deal easily with linear differential equations with constant coefficients, and an awareness of how many important physical phenomena, such as resonance, criticality, and so on, can be understood both qualitatively and quantitatively in terms of the behavior of the solutions of differential equations as functions of both time and system parameters.

3. An ability to use power-series expansions to obtain analytic and computational behavior of the solutions of linear and nonlinear differential equations. In addition, the reader should learn to appreciate the use of perturbation expansions in terms of various parameters in the equation.

4. An understanding of the basic methods used to obtain numerical solutions of differential equations using both desk and digital computers. We assume that he has learned the rudiments of FORTRAN programming prior to taking this course, or simultaneously. However, this ability is required only in one part of the text, Chapter 4. The remainder of the book can be read by the computerless student.

5. An appreciation for the use of successive approximations as an all-purpose tool for obtaining analytic and numerical solutions of differential equations, and as a technique for establishing existence theorems.

The foregoing list constitutes a rather tall order, and it is not at all clear that all these ideas and techniques can be presented in a one-semester course. It is far more important to present half of it in a careful manner than to cover it all hurriedly and thereby fill the student with a number of recipes in a "cookbook course."

It is essential that the reader understand the importance of rigor. This does not mean that every result has to be established rigorously, nor that epsilons and deltas bedeck every page. We feel that it is sufficient to point out the many pitfalls of formalism, to make him aware of when a formal method is being employed and when a result which can be established rigorously is being used. Much depends, of course, on the caliber of the class and on the personal predilections of the instructor. No book is ever intended to be adhered to rigidly.

The book is divided into six chapters, each designed to furnish one of the skills sketched above. In Chapter 1, a number of physical processes drawn from different areas of science are formulated in terms of differential equations. Chapter 2 treats in detail the analytic solution of linear differential equations of the second order with constant coefficients, covering both initial-value and two-point boundary-value problems. Power series both in independent variables and in system parameters constitutes the theme of Chapter 3. We show by means of examples how easy it is to use power-series expansions to obtain numerical solutions. In Chapter 4, we present a different technique—the use of difference approximations—for computational solutions. In particular, we emphasize the flexibility of a digital-computer solution which can employ an adaptive program whereby the equation itself determines the length of the computing time. We also could not resist including a small section devoted to the identification of systems. Here the simple idea is that any device, such as a digital computer, that can perform calculations with enormous rapidity can be used to directly determine the structure of a system on the basis of observations. The determination of structures is of course one of the basic problems of science.

Chapter 5 is devoted to the study of systems of differential equations. We do not pursue the subject in any depth since vector notation and matrix theory are required for any meaningful presentation. We do take advantage of the numerical difficulties involved in solving initial-value problems of high-order systems to motivate the use of the Laplace transform. At the end of this chapter, we indicate briefly again how the ability to solve huge systems of ordinary differential equations enables us to deal in a reasonably simple fashion with the numerical solution of many types of partial differential equations.

The concluding chapter contains a proof of uniqueness of solution, a proof of the existence of solution using the powerful method of successive approximations, and a discussion of the convergence of difference approximations.

The examples at the end of each section, and again at the end of each chapter, have been carefully chosen, on the whole, to illustrate a number of different ideas and methods. We are not enthusiastic about the usual textbook practice of including many problems of exactly the same type, differing only in the values of the parameters. We believe that a careful discussion of a single problem is far more important than routine drill. The instructor who disagrees can, of course, easily construct his own drill problems.

At the end of each chapter there are a number of references to more advanced results for the reader who wishes to follow a particular direction, together with comments which might have distracted the reader if included in the text.

We hope that we have communicated some of the fun that a mathematician can have in applying the theory of differential equations to various categories of problems. With the aid of the great scientific equalizer, the digital computer, the researcher of even moderate mathematical ability can formulate his own mathematical models, test his hypotheses, and roam at will throughout the scientific domain. This book is intended to be a passport. There has never been a more exciting time in intellectual history than the present to take this journey.

At this point, I wish to thank a number of friends and colleagues for their help in the preparation of this book. T. J. Higgins, M. Sedlar, and R. Varga read through the manuscript with great care and made a number of valuable suggestions which materially improved the text. J. Buell prepared the FORTRAN programs and contributed substantially to the numerical results of Chapter Four. For the preparation of the manuscript in her usual peerless fashion, I express my appreciation to Jeanette Blood.

Los Angeles, California
August 1967

Richard Bellman

Contents

Chapter 1 The Origins of Differential Equations

Chapter 2 Linear Differential Equations and Their Solutions in Compact Forms

Chapter 3 Power-Series Solutions

CHAPTER 1

The Origins of Differential Equations

1.1 INTRODUCTION

In this chapter, we wish to introduce the reader to some simple but important types of differential equations which arise most naturally in the various fields of science. Our examples will be drawn from physics, engineering, chemistry, biology, economics, and operations research, thus demonstrating rather conclusively that one of the principal languages of science is that of differential equations.

We first show by means of a number of illustrative examples how easy it is to translate physical phenomena into mathematical equations, using the elegant and versatile tool of calculus. Some of these equations can be completely solved using only the most elementary techniques; others require for their analytic or computational solution the techniques we introduce in later chapters.

By discussing some of these problems in detail, we hope to make clear what is meant by the term "solution," and what we can reasonably expect from a theory of differential equations.

1.2 WHAT IS A DIFFERENTIAL EQUATION?

In calculus we encounter the concept of the derivative of a function. If $y(x)$ is the function, we denote the derivative by the symbols dy/dx, $y'(x)$, or simply y' when there is no possible ambiguity about the independent variable. Similarly, the second derivative is denoted by d^2y/dx^2, $y''(x)$, or y''; the nth derivative by d^ny/dx^n, $y^{(n)}(x)$, or $y^{(n)}$.

A differential equation is quite simply an equation connecting a function $y(x)$ and some of its derivatives. The simplest type of differential equation is that encountered in integral calculus, namely

$$\frac{dy}{dx} = g(x). \tag{2.1}$$

In words: What is the function $y(x)$ whose derivative is $g(x)$?

Most differential equations are more difficult to treat, and correspondingly more interesting to study. We shall consider equations such as

$$\frac{d^2y}{dx^2} + \sin y = 0, \tag{2.2a}$$

arising in the study of the motion of the pendulum,

$$\frac{d^2u}{dt^2} + a\frac{du}{dt} + bu = c\cos\omega t, \tag{2.2b}$$

arising in the analysis of simple electric circuits, and

$$\frac{d^2y}{dx^2} + xy = 0, \tag{2.2c}$$

arising in astronomical investigations, as well as in many other places.

Sometimes we consider y to be a function of x; sometimes u a function of t, and so on. In some cases, the notation has physical significance, e.g., t denotes time, x denotes distance; in other cases, it is merely a matter of choosing some convenient letters. We have alternated between the use of t and x in our examples to accustom the reader to the sight of equations in different variables.

Once we get the idea of what a differential equation is, it is clear that we can write down as many different kinds as we please. On which ones, then, should we focus our attention? In an introductory course such as this, we feel that it is essential to begin with the simplest and most important categories of equations—covering significant areas of mathematics and scientific phenomena which illustrate fundamental analytic techniques.

1.3 WHAT IS A SOLUTION?

Algebra teaches us what is meant by the injunction "Determine all solutions of

$$x^2 - 3x + 2 = 0." \tag{3.1}$$

We are asked to find all numbers, real or complex, with the property that substitution in (3.1) yields an identity.

Similarly, given a differential equation such as

$$\frac{d^2u}{dt^2} + u = 0, \tag{3.2}$$

we wish to find all functions $u(t)$ which make this expression an identity. Whereas the algebraic equation in (3.1) has exactly two solutions, $x = 1$ and $x = 2$, we see that there are many functions satisfying (3.2). Thus $u = \cos t$ is a solution, and so also are $u = \sin t$ and $u = 3 \cos t - \sin t$; and generally any function of the form $u = c_1 \cos t + c_2 \sin t$, where c_1 and c_2 are constants, satisfies (3.2).

At first, this proliferation of solutions is disconcerting. As we continue, however, we shall see that this freedom in the choice of solutions is exactly what we require in the treatment of physical processes. This will be illustrated by a number of examples in this and the following chapters.

The differential equation (3.2), which is a particular case of Eq. (2.2b), is an example of an equation which possesses solutions that can be expressed in terms of the familiar functions of analysis. This property is quite exceptional and should not usually be anticipated. The great majority of the equations which arise in theory and in applications do not possess this convenient property. For example, both of the Eqs. (2.2a) and (2.2c) give rise to new types of functions not found in elementary calculus.

We cannot, of course, accept this as a reason for ending our study of these equations. The questions still remain: How do we obtain analytic and numerical solutions for the general case? How do we recognize which equations have simple solutions, and what do we mean by a simple solution?

These are extremely interesting and important questions which will never be completely answered. New differential equations are constantly appearing in scientific research, requiring new mathematical techniques for their solution and introducing new and important classes of functions. Furthermore, let us emphasize that the development of new technological devices, such as analog computers, digital computers, and hybrid computers, constantly alters the whole concept of "simple solution." We shall discuss this important point again in Chapter 4, which is devoted to the problem of numerical solution.

Our aim in this introductory volume is to present a few basic analytic and computational ideas which will enable the reader to use and understand differential equations in his work and to obtain, with the use of a digital computer, numerical solutions of complex differential equations arising in the detailed study of economic, engineering, physical, and biomedical phenomena. Hopefully, he will be stimulated to continue with intermediate and advanced work either as a professional mathematician or as a means of engaging in more sophisticated scientific research.

1.4 WHERE DO DIFFERENTIAL EQUATIONS ARISE?

The theory of differential equations is extremely rich in results and fertile in ideas. So far as we can see, it shows no sign of exhaustion despite the hundreds of mathematicians studying problems in this field. It remains a fundamental domain of analysis. To the nonmathematician, however, differential equations are important because they constitute the language of modern science. There are alternative descriptions of the real world, but none as flexible, as versatile, as easy to manipulate analytically, and as convenient for modern computers.

We shall show, by means of examples across the spectrum of science, that we can simply and quickly study important classes of phenomena by means of differential equations. In the sections that follow, we shall show how to convert verbal description of various kinds of physical processes into differential equations. Some equations of a particularly simple type will be solved immediately and discussed. In Chapter 2, a number of the equations which escape solution by these direct techniques will be considered, together with a number of further applications from various fields. In Chapter 3, we shall discuss two of the most powerful analytic methods for the solution of otherwise intractable equations: power-series expansions and perturbation techniques, both in elementary guise. In Chapter 4, we shall present a general method for using a digital computer to obtain numerical solutions of differential equations and, once again, illustrate the technique by means of a number of applications.

This volume is, as we indicated in the Preface, problem and method oriented. This approach has the advantage of enabling us to readily make a crucial decision: Which differential equations to study first.

There are so many different kinds of differential equations that one can write down. Clearly, we should like to begin with those equations which possess interesting characteristics and are amenable to analytic solutions. How can we be sure of our equations in advance?

There is a very simple test: "A meaningful scientific origin of an equation guarantees an interesting problem and a meaningful mathematical solution." This simple principle has been repeatedly and successfully invoked over the past three hundred years. And, as a matter of fact, it seems to be the only safe policy to follow. Up to the present time, man's imagination has not been able to compete with the ability of the physical world in the conjuring up of new and intriguing problems to study.

A more intensive and extensive study of natural phenomena leads to more complicated types of equations: partial differential, integral, differential-difference, and equations with even more frightening and exotic names. We mention these to reiterate our point that the physical world is a continuing and endless source of interesting mathematics. The reader need not fear that there will be nothing left to investigate by the time he finishes his training. If anything, there will be very much more.

1.5 MAXIMUM ALTITUDE (1)

Let us begin our task of translating various aspects of physical processes into differential equations with the following familiar problem: If we throw a ball straight up, how high will it go?

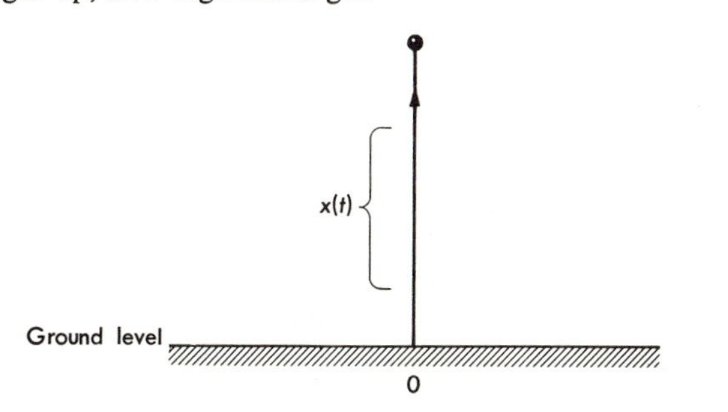

Figure 1.1

There are several ways of treating this question. Let us consider the conventional approach under the usual assumptions that the ball is a point particle, that the earth is flat, and that there is no resistance due to air or wind.

Consider Fig. 1.1. Let

(5.1) $x(t)$ = the height of the ball at time t, supposing that the ball is thrown straight up at time $t = 0$ with velocity v.

If the ball is subject only to the force of gravity, an acceleration downward, we obtain the equation

$$\frac{d^2x}{dt^2} = -g, \tag{5.2}$$

where $g = 32$ ft/sec^2, recalling that the second derivative represents acceleration. We have taken the unit of mass to be 1. This is a simple trajectory process.

Since we started at ground level at time zero, we have the condition

$$x(0) = 0, \tag{5.3}$$

called an initial condition because it provides information about the solution at time zero; and since the ball was thrown upward with velocity v, we have a second initial condition:

$$x'(0) = v. \tag{5.4}$$

Let us now show that these data, together with Eq. (5.2), determine the path of the ball. Integrating both sides of (5.2) from 0 to t, we have

$$\int_0^t \frac{d^2x}{dt_1^2}\,dt_1 = -\int_0^t g\,dt_1, \qquad \frac{dx}{dt}\bigg]_0^t = -gt. \tag{5.5}$$

Since (5.4) holds, the second equation of (5.5) yields

$$\frac{dx}{dt} - v = -gt, \qquad \text{or} \qquad \frac{dx}{dt} = v - gt. \tag{5.6}$$

Hence we have deduced an expression for the velocity at any time that the ball is in the air.

At the moment the maximum height is attained, the velocity must be zero. Hence this height is reached at time v/g. To find the altitude at v/g, we repeat the foregoing procedure, starting with Eq. (5.6). We obtain

$$\int_0^t \frac{dx}{dt_1}\,dt_1 = \int_0^t (v - gt_1)\,dt_1, \qquad x]_0^t = vt - \frac{gt^2}{2}. \tag{5.7}$$

Using Eq. (5.3), we find that

$$x = vt - gt^2/2. \tag{5.8}$$

At time $t = v/g$ we have

$$x_{\max} = v\left(\frac{v}{g}\right) - \frac{g}{2}\left(\frac{v}{g}\right)^2 = \frac{v^2}{2g}. \tag{5.9}$$

EXERCISES

---

1. Given $h < v^2/2g$, find the two times at which the height of the ball is h.

2. Suppose that a ball is dropped from a height h. How long will it take to reach the ground?

3. Suppose that at initial height h a ball is thrown up with a velocity v. How long will it take to reach the maximum height and how long to reach the ground?

1.6 MAXIMUM ALTITUDE (2)

The foregoing analysis answers the original question, but tells us perhaps more than we wanted to know. Suppose we only wished to determine the maximum altitude, not the whole time history of the trajectory process.

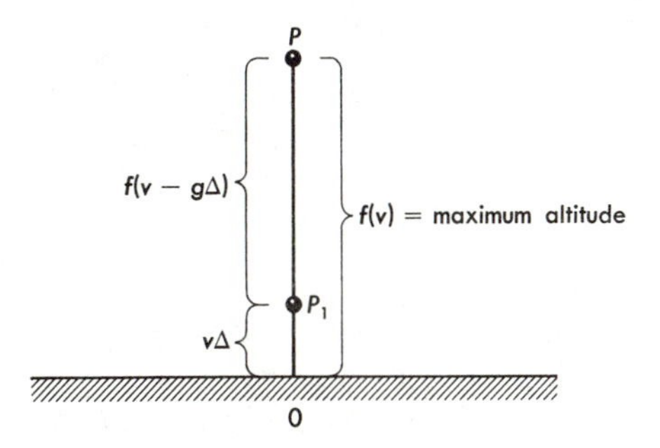

Figure 1.2

Let us adopt a different approach aimed solely at the determination of the maximum altitude. We begin with the observation that the maximum altitude depends on the initial velocity v. Hence we introduce the function

(6.1) $f(v)$ = the maximum altitude attained starting with initial velocity v.

Let the ball travel upward for a short time Δ. As a matter of fact, Δ is to be taken so small that Δ^2 can be safely ignored in any equation involving Δ^2. In the older literature, Δ was called an "infinitesimal," a term that has now come into disrepute in some quarters. It is, however, a convenient term when properly applied, and so we shall use it occasionally below. During this brief time, the ball gains altitude, $v\Delta$, and loses velocity $g\Delta$ due to the acceleration of gravity directed downward. The velocity at P_1 is thus $v - g\Delta$ (see Fig. 1.2). These values are, of course, approximations which are sufficiently accurate since Δ is, by assumption, an "infinitesimal." If OP is the maximum altitude $f(v)$, we see that P_1P is the maximum altitude gained starting from P_1 with velocity $v - g\Delta$. Thus, to our degree of approximation,

$$f(v) = v\Delta + f(v - g\Delta). \tag{6.2}$$

Expanding the right-hand side by means of Taylor's series, we have

$$f(v) = v\Delta + f(v) - g\,\Delta f'(v) + \text{terms of order } \Delta^2. \tag{6.3}$$

As agreed above, we can ignore the terms of order Δ^2. Canceling the $f(v)$

term and dividing through by Δ, we have the relation

$$0 = v - gf'(v) + \text{terms of order } \Delta. \tag{6.4}$$

In the limit as $\Delta \to 0$, we end up with the equation

$$0 = v - gf'(v), \qquad \text{or} \qquad v/g = f'(v). \tag{6.5}$$

The initial condition is clearly $f(0) = 0$, since zero initial velocity produces no increase in altitude. Hence, proceeding as before, we have

$$f(v) = v^2/2g, \tag{6.6}$$

which agrees with our previous result.

In deriving the foregoing equation (6.5) our basic assumption was that $f(v)$ was a function which possessed a Taylor expansion in v, or at least a second derivative for $v \geq 0$. In the above example, this is easy to verify by using our knowledge of the explicit form of $f(v)$ provided by Eq. (5.9).

In general, verification of the above-mentioned property requires a careful investigation of one type of another. In all that follows in this chapter, we shall tacitly assume that the functions we consider are sufficiently differentiable. Sometimes we can establish this fact directly. Sometimes we proceed boldly and define our functions as the solutions of the equations we have derived. In the latter event, there then remains the test of comparing functions obtained with the results of observation and experiment. If the agreement is satisfactory, we keep the equation as a "law of nature." If not, we revise some of our concepts.

EXERCISE

1. Suppose that we wish to determine the time required for the ball to attain maximum altitude, using the foregoing approach. Let $h(v)$ denote this time. Show that if Δ is an infinitesimal, then $h(v) = \Delta + h(v - g\Delta)$, and thus $h'(v) = 1/g$. Using the condition $h(0) = 0$, show that $h(v) = v/g$.

1.7 GROWTH OF A POPULATION (1)

Growth processes are of great importance, whether they be the growth of a human population, of a colony of cells in the laboratory, or in a human organ, or of a school of fish. Growth processes were first investigated using the tool of differential equations by the Italian mathematician V. Volterra and the American mathematician A. J. Lotka, and subsequently by many others. Let us see how one can proceed.

Let $N(t)$ denote the number of individuals in a population at time t. This number is of course an integer, and it will change over time by integral amounts. Nonetheless, if $N(t)$ is large, we can reasonably consider it to be a continuous variable in the sense that the change in $N(t)$ over a small time interval will be small compared to the magnitude of $N(t)$.

We begin with a plausible assumption that the rate of change of the population at any time is directly proportional to the size of the population at that time. The mathematical translation of this hypothesis is the equation

$$\frac{dN}{dt} = kN, \tag{7.1}$$

where k is the constant of proportionality. What does an equation of this type imply as far as the time history of the growth process is concerned?

Since this is a simple model of a growth process, we look for a simple function which satisfies Eq. (7.1). We recall that

$$\frac{d}{dt}(e^{bt}) = be^{bt} \tag{7.2}$$

for any constant b. Comparing (7.1) and (7.2), we see that $N(t) = e^{kt}$ is a solution of (7.1). Moreover, $N(t) = ce^{kt}$ is a solution of the equation for any constant c. This knowledge is useful, since by choosing c properly we can fit an initial condition, namely that $N(0)$ is a given quantity N_0. Hence, as a candidate for the solution of (7.1), we have

$$N(t) = N_0 e^{kt}. \tag{7.3}$$

Is (7.3) the only candidate? Suppose that we had momentarily forgotten (7.2). Since $N(t)$ is the size of the population at time t, and thus never zero, we are tempted to proceed as follows. Starting with (7.1), write

$$\frac{1}{N}\frac{dN}{dt} = k \tag{7.4}$$

and integrate both sides between 0 and t. We thus obtain

(7.5)

$$\int_0^t \frac{1}{N}\frac{dN}{dt_1}\,dt_1 = \int_0^t k\,dt_1, \qquad \log N]_0^t = kt, \qquad \log N(t) - \log N_0 = kt,$$

whence (7.3) follows.

Are there other solutions which become zero at various times—extraneous solutions so far as the original growth process is concerned? It turns out that there are none, but the proof requires a different artifice which we shall employ in the next chapter. In the meantime, let us proceed with the discussion of the solution just given.

1.8 THE INVERSE PROBLEM

In many scientific investigations, the use of mathematics is dictated not so much by a desire to solve equations as by the desire to test hypotheses. Suppose that a population of cells is measured at various times, yielding the graph shown in Fig. 1.3. Is it true that the rate of growth is directly proportional to the size of the population? By means of the mathematical model of the growth process we have constructed we can convert this scientific question into a curve-fitting problem. Given the data, the values of $N(t)$, we wish to determine a value of k with the property that experimentally determined values of $N(t)$ are close to the values obtained from the analytic solution.

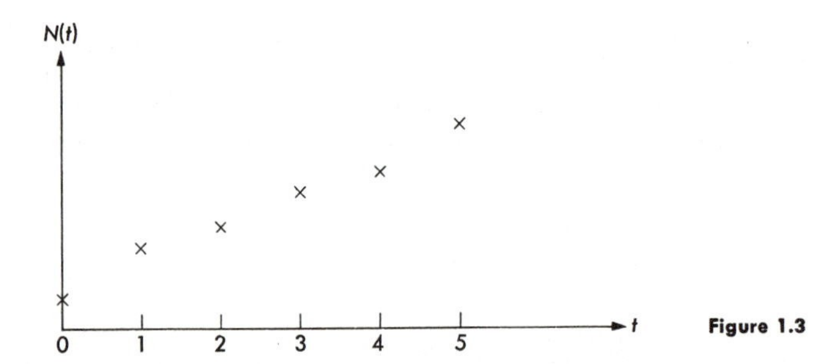

Figure 1.3

This problem of determining the basic equation, or parameters in the equation, given the observed behavior, is often called the "inverse problem." It is one of the fundamental problems of science, since theories stand or fall on the basis of comparison between prediction and observation.

In the present case, the determination of k is quite simple. If $N(t)$ has the form of (7.3), then

(8.1) $$\log N(t) = \log N_0 + kt.$$

Hence $\log N(t)$ plotted against t should be a straight line with slope k, as in Fig. 1.4. If, as a result of either experimental error or deviations from ideal behavior, the points are not exactly on a straight line, we can obtain an

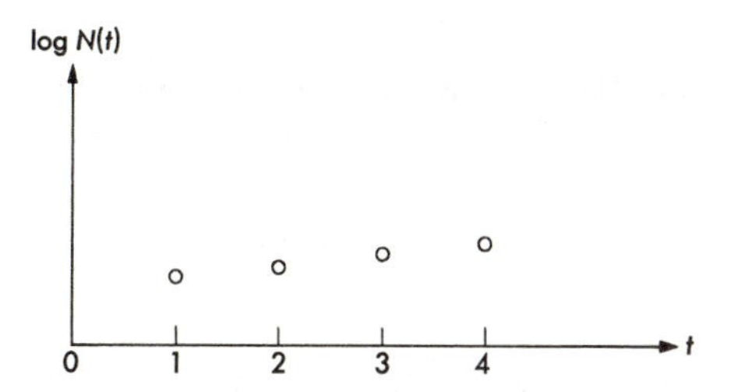

Figure 1.4

approximate value for k by some mean-square procedure, or by other approximation procedures.

Thus, for example, if we let $E(t)$ denote the observed size of the population at time t and if we have measurements at times $t_1, t_2, \ldots, t_N$, we can ask that $\log N_0$ and k be determined by the condition that the expression

$$D_N = \sum_{i=1}^{N} [\log E(t_i) - \log N_0 - kt_i]^2 \tag{8.2}$$

be minimized with respect to these variables.

EXERCISE

1. Consider D_N as a quadratic expression in $\log N_0$ and k and, using calculus, obtain the values which minimize D_N.

1.9 GROWTH OF A POPULATION (2)

The growth curve of many populations possesses a part that corresponds very closely to the exponential growth predicted above. If, however, we examine the growth over a longer period of time, we obtain data of the type in Fig. 1.5. There is a very definite leveling-off effect.

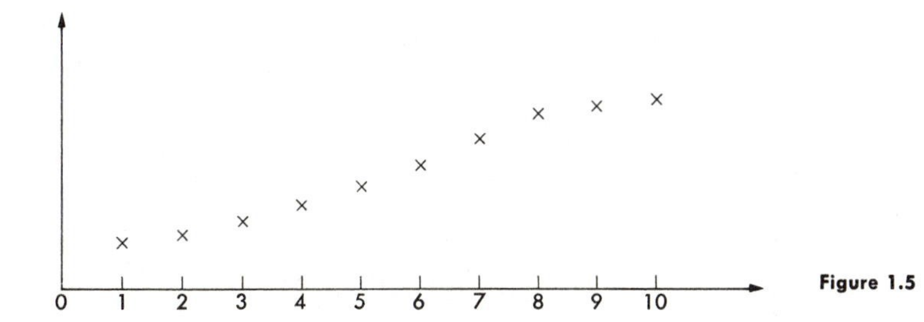

Figure 1.5

How do we attempt to explain this effect? We can consider the change in the size of a population as a consequence of two phenomena, birth and death. Hence, if we let b denote the birth rate and m the death rate, we can write

$$\frac{dN}{dt} = (b - m)N. \tag{9.1}$$

So far there is nothing new. Equation (9.1) merely elucidates the meaning of the constant k. Let us now observe that it is plausible that b and m may themselves depend on the size of the population. If, for example, we consider that the food supply is limited, it is reasonable to expect that m will increase as N increases. Let us take the simplest case of direct proportionality. Suppose that $m = m_1 N$ so that (9.1) becomes

$$\frac{dN}{dt} = (b - m_1 N)N. \tag{9.2}$$

Proceeding as before, we write

$$\frac{1}{(b - m_1 N)N} \frac{dN}{dt} = 1 \tag{9.3}$$

(hoping that $b - m_1 N$ remains positive), and integrate both sides between 0 and t:

$$\int_0^t \left[\frac{1}{(b - m_1 N)N}\right] \frac{dN}{dt_1}\, dt_1 = \int_0^t dt_1 = t. \tag{9.4}$$

Using a decomposition into partial fractions,

$$\frac{1}{(b - m_1 N)N} = \frac{1}{b}\left(\frac{m_1}{b - m_1 N} + \frac{1}{N}\right), \tag{9.5}$$

we see that (9.4) yields

$$\frac{1}{b}[-\log (b - m_1 N) + \log N]_0^t = t, \tag{9.6}$$

or

$$\log\left(\frac{N}{b - m_1 N}\right) - \log\left(\frac{N_0}{b - m_1 N_0}\right) = bt. \tag{9.7}$$

Thus

$$\frac{N}{b - m_1 N} = \frac{N_0}{b - m_1 N_0}\, e^{bt}, \tag{9.8}$$

or, finally, with a small amount of algebra,

$$N = \frac{[bN_0/(b - m_1N_0)]e^{bt}}{1 + [m_1N_0/(b - m_1N_0)]e^{bt}}. \tag{9.9}$$

From Eqs. (9.8) and (9.9) we see that our procedure is self-consistent in the sense that $b - m_1N$ remains positive if it is positive at time $t = 0$.

If we graph N as a function of t, we find precisely the desired behavior: exponential growth for small time and a leveling-off as t increases. We see also how to estimate the parameters b and m_1. The parameter b can be determined from the original exponential growth, and the ratio b/m_1 is the final limiting size of the population.

Turning to Eq. (9.2), the first property is a consequence of writing

$$(b - m_1N)N = bN - m_1N^2 \tag{9.10}$$

and ignoring the term m_1N^2 in favor of bN for N small. Thus, we obtain the approximate equation

$$\frac{dN}{dt} \cong bN, \tag{9.11}$$

(read dN/dt is approximately equal to bN) and thus, presumably, the relation

$$N \cong N_0e^{bt}. \tag{9.12}$$

The study of approximate relations such as (9.11) is part of the *stability theory* of differential equations, and belongs to the advanced theory of differential equations. Consequently, we shall not consider questions of this nature here. As the reader may imagine, it is a theory of major importance in the applications of differential equations to the scientific world.

The limiting value b/m_1 is obtained by looking at the right-hand side of (9.2) and asking for "steady-state" values, i.e., values of N for which $dN/dt = 0$. There are two values, the uninteresting one, $N = 0$, and the stated value b/m_1, which we expect to be the limiting size of the population. The proof that this is indeed the case is readily seen using (9.9).

1.10 POPULATION EXPLOSION

Suppose we had assumed that b, the birth rate, was directly proportional to the size of the population. Then, instead of (9.2), we would have

$$\frac{dN}{dt} = (b_1N - m)N. \tag{10.1}$$

Proceeding as before, we have

$$\int_0^t \left[\frac{1}{(b_1 N - m)N}\right]\frac{dN}{dt_1}\,dt_1 = t, \tag{10.2}$$

$$\int_0^t \frac{1}{m}\left[\frac{b_1}{b_1 N - m} - \frac{1}{N}\right]dt_1 = t,$$

$$[\log (b_1 N - m) - \log N]_0^t = mt,$$

$$\frac{b_1 N - m}{N} = \frac{b_1 N_0 - m}{N_0} e^{mt},$$

and thus, finally,

$$N = \frac{m}{b_1 - [(b_1 N_0 - m)/N_0]e^{mt}}. \tag{10.3}$$

Let us suppose that N_0 is sufficiently large so that $b_1 N_0 - m > 0$. We now observe the following phenomenon. As t increases, the function $N(t)$ increases steadily until the point where

$$b_1 - \left(\frac{b_1 N_0 - m}{N_0}\right) e^{mt} = 0. \tag{10.4}$$

(See Fig. 1.6.) We can consider this to be a mathematical model of "population explosion," or, alternatively, of a nuclear fission process. In Section 1.18 we shall discuss this in more detail.

1.11 GEOMETRIC PROBLEMS

Many simple geometric considerations lead to problems concerning differential equations. For example, if we ask for a curve with constant slope, we obtain the differential equation

$$y' = m, \tag{11.1}$$

a not very exciting equation. If, however, we ask for a curve with constant curvature, we are led to the more complex equation

$$\frac{y''}{(1 + y'^2)^{3/2}} = k. \tag{11.2}$$

Similarly, as we show in the exercises below, other geometric properties lead to other differential equations.

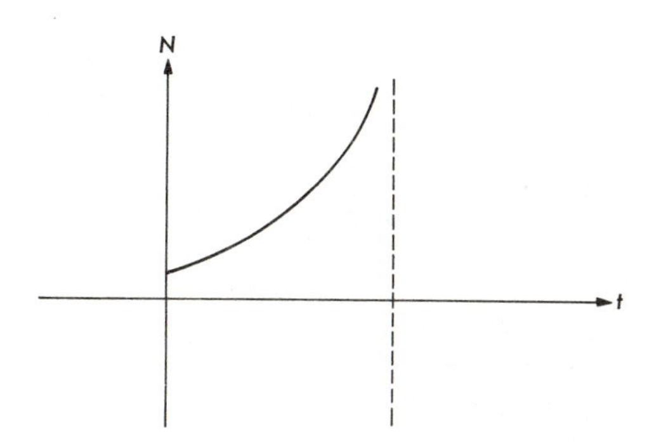

Figure 1.6

EXERCISES

1. Show that

$$y = mx + n$$

(m and n are constants) is a solution of Eq. (11.1), and also of (11.2) if $k = 0$.

2. Show that

$$(y - a)^2 + (x - b)^2 = 1/k^2$$

defines a solution of (11.2) for arbitrary constants a and b given that $k \neq 0$.

3. A parabolic reflector has the property that a light source placed at its focus produces a parallel beam, or, conversely, parallel rays converge at the focus (Fig. 1.7). Assuming that reflection of light from a curve is determined by the usual laws of reflection for the tangent to the curve at the point of incidence (angle of incidence equals angle of reflection), use the above property to determine a differential equation for the parabola.

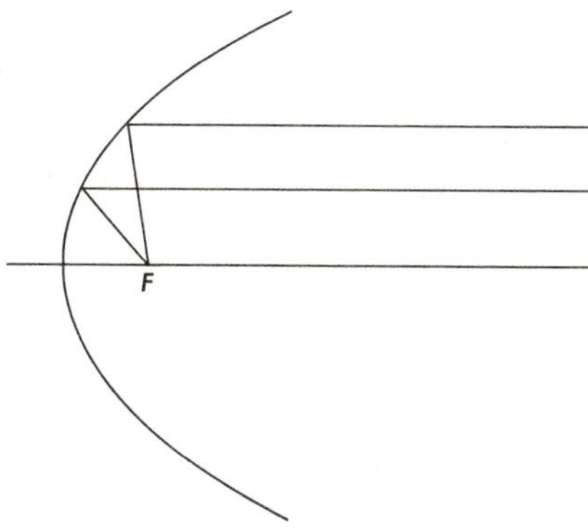

Figure 1.7

4. Let $r = r(\theta)$ be the equation of a curve in polar coordinates. Obtain a differential equation for $r(\theta)$ using the property that the arc length from $\theta = 0$ to $\theta = \theta_1$ is proportional to θ_1 for any $\theta_1 \geq 0$.

5. Let $u(x)$ be a function with the property that the area under the curve between any two points, a, b with $a < b$, is directly proportional to the difference of the functional values at a and b. Obtain a differential equation for $u(x)$. [*Hint:* The first statement is equivalent to the relation $\int_a^b u(x)\,dx = k\big(u(b) - u(a)\big)$. This is an *integral equation*. Consider b as a variable and differentiate to obtain the differential equation $u(b) = ku'(b)$ valid for all b.]

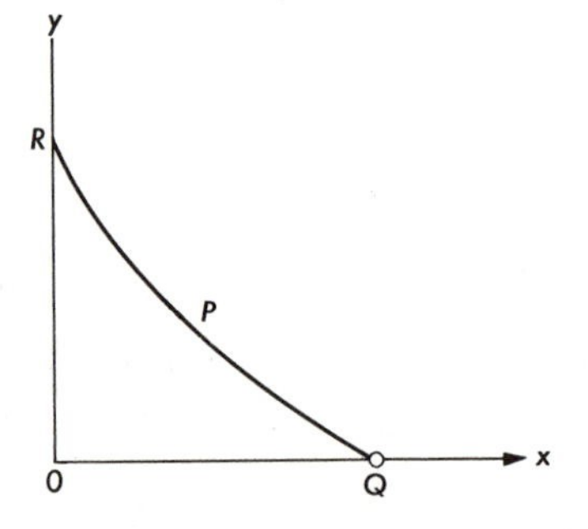

Figure 1.8

6. Let $y = y(x)$ be a curve, as indicated in Fig. 1.8, with the property that particles placed at any points P along the curve and falling under the influence of gravity along the curve with no friction, will land at Q at the same time. What equation do we obtain by using this property? (This problem is included to reinforce the point that quite simple and natural questions can lead to equations which are not differential equations.)

   The equation obtained in this fashion is one of the famous integral equations in analysis, the Abel integral equation.

1.12 CURVES OF PURSUIT

Many interesting classes of differential equations arise in the study of trajectory processes and in pursuit processes where one object is constrained to follow another. A famous one which we shall discuss numerically in Chapter 4 reads in this fashion. A rabbit, initially at $(a, 0)$, travels along the x-axis in the direction of increasing x with constant velocity v_R. A dog, initially at $(0, b)$, follows the rabbit with constant velocity v_D, with the direc-

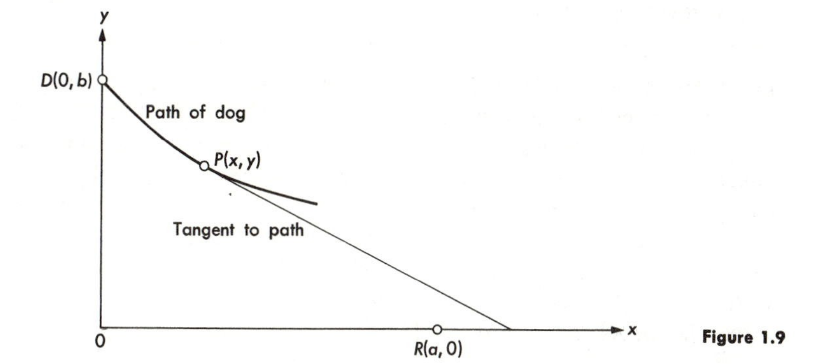

Figure 1.9

tion of motion pointed toward the rabbit at all times. (See Fig. 1.9.) We are asked to determine the path of the dog.

The fact that the dog points at the rabbit implies that, at any time t, the tangent to the curve traced out by the dog's motion passes through the position of the rabbit at that time (Fig. 1.9).

Let the equation describing the path of the dog as a function of time be

$$x = x(t), \qquad y = y(t). \tag{12.1}$$

The equation of the tangent at the point $(x(t), y(t))$ is

$$y - y(t) = \frac{y'(t)}{x'(t)}[x - x(t)]. \tag{12.2}$$

Since this passes through the position of the rabbit at time t, namely $(a + v_R t, 0)$, we have the relation

$$-y(t) = \frac{y'(t)}{x'(t)}[a + v_R t - x(t)]. \tag{12.3}$$

To obtain another relation, we use the fact that the magnitude of the velocity of the dog is constant and equal to v_D. Thus,

$$v_D = \left[\left(\frac{dx}{dt}\right)^2 + \left(\frac{dy}{dt}\right)^2\right]^{1/2}. \tag{12.4}$$

We thus obtain a set of simultaneous differential equations for x and y as functions of t. The initial conditions are

$$x(0) = 0, \qquad y(0) = b. \tag{12.5}$$

With some ingenuity, the above equations can be integrated in terms of elementary functions of analysis.

1.13 ELECTRIC CIRCUITS

Let us now consider one of the most important technological systems in our society: an electric circuit consisting of a resistor, a capacitor, and an inductor in series with a voltage source (see Fig. 1.10). The most important example of such a system is one in which $E = E_0 \cos \omega t$, with the special case $E = E_0$, corresponding to $\omega = 0$. We shall assume here that the voltage is constant. We wish to determine the current in the system as a function of time. If we denote the current by i, short for $i(t)$, we know that at any time t the voltage across the resistor is iR, that across the inductor is $L\,di/dt$, and that across the capacitor is $\int_0^t i\,dt_1/c$. Hence, using Kirchhoff's

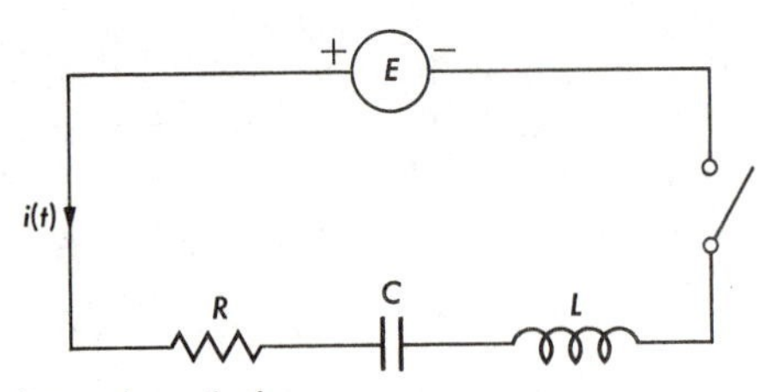

Figure 1.10

voltage law, we have the relation

$$L\frac{di}{dt} + Ri + \int_0^t \frac{i\,dt_1}{C} = E_0. \tag{13.1}$$

Observe that, because of the presence of the integral term, this is *not* a differential equation. To obtain an equivalent differential equation, we differentiate both sides of (13.1):

$$L\frac{d^2i}{dt^2} + R\frac{di}{dt} + \frac{i}{C} = 0. \tag{13.2}$$

What are the initial conditions? To begin with, we suppose that the switch is closed in the circuit at time $t = 0$. At that time there is certainly no current flow. Hence

$$i(0) = 0. \tag{13.3}$$

Secondly, turning to (13.1) and setting $t = 0$, we obtain

$$L\frac{di}{dt} + Ri = E_0. \tag{13.4}$$

Since $i(0) = 0$, this equation reduces to

$$Li'(0) = E_0. \tag{13.5}$$

In the following chapter we shall show that these two conditions, (13.3) and (13.5), determine the time history of the current, and we will discuss particular circuits and various phenomena associated with them.

EXERCISES

1. Show how the equation $u(t) + \int_0^t e^{a(t-t_1)}u(t_1)\,dt_1 = k$ can be reduced to a differential equation and obtain an initial condition for this equation.
2. Show how the integral equation

$$u(t) + b_1 \int_0^t a^{-a_1(t-t_1)}u(t_1)\,dt_1 + b_2 \int_0^t e^{-a_2(t-t_1)}u(t_1)\,dt_1 = k$$

can be reduced to a differential equation and obtain initial conditions for the resulting equation.
3. Show how the integral equation

$$u(x) + b \int_0^1 e^{-a|x-y|}u(y)\,dy = k$$

can be reduced to a differential equation and obtain boundary conditions at $x = 0$ and $x = 1$.

1.14 ANALOG COMPUTERS

We have shown that every *LRC*-circuit of the type discussed in Section 1.13 gives rise to a linear differential equation of the kind appearing in (13.2). Conversely, every differential equation of this type with nonnegative coefficients can be considered as arising from an *LRC*-circuit of the foregoing kind.

Could we not then use this correspondence to obtain the numerical solution for Eq. (13.2)? Starting with the equation, we can build a circuit (see Fig. 1.11) containing a switch, an ammeter to read current, and adjustable resistors, capacitors, and inductors. Setting these devices at the correct values, we can close the switch and *read* the solution, $i(t)$, on the ammeter. Putting a voltmeter across the inductor, we can read $L\,di/dt$ at the same time.

This is the basic idea of the analog computer, one of the important ideas in modern science. What we are exploiting is the fact that similar differential

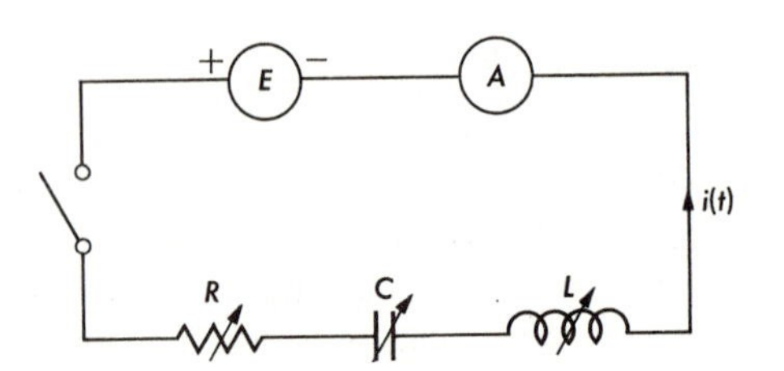

Figure 1.11

equations can arise in many physical areas. We are then free to use whichever domain is most convenient for our purposes in constructing an analog system.

The device is constructed with adjustable elements so that we can use it over and over again to solve equations of this general type. Although we shall focus our attention on the digital computer in Chapter 4, we do not wish to minimize the importance of the concept of the analog computer.

As a matter of fact, the new and exciting development in the computer field is the emergence of the hybrid computer, containing both digital and analog capabilities. All of this means that the chapter on numerical analysis will require extensive rewriting in a very few years. We are not particularly disturbed at this prospect, since our aim is to convey the spirit of the modern approach to differential equations and to science involving the use of computers, without dwelling overly on the technical aspects.

1.15 MATHEMATICAL ECONOMICS: MAXIMIZATION OF PROFIT

The field of mathematical economics is a prolific source of problems involving differential equations. Consider, for example, the question of operating a business so as to maximize the total profit over a given time period.

Let the total dollar value of an enterprise at time t be denoted by $u(t)$ and suppose that the rate of increase is directly proportional to the value. We thus obtain the familiar equation

$$\frac{du}{dt} = au, \qquad u(0) = c, \tag{15.1}$$

where c is the value at some initial time. This yields exponential growth $u = ce^{at}$.

Although growth of the firm is quite satisfactory, the principal objective in its management is profit. Suppose that it is decided to reduce the dollar value of the business at a rate equal to a fixed fraction k of the value. If

$p(t)$ denotes the total profit made over the interval $[0, t]$, we then have

$$\frac{dp}{dt} = ku, \qquad p(0) = 0; \tag{15.2}$$

and in place of Eq. (15.1) we have

$$\frac{du}{dt} = a(1 - k)u, \qquad u(0) = c. \tag{15.3}$$

Then

$$u = ce^{a(1-k)t} \tag{15.4}$$

and

$$p = k \int_0^t [ce^{a(1-k)t_1}]\, dt_1 = \frac{kc}{a(1 - k)} [e^{a(1-k)t} - 1]. \tag{15.5}$$

How does the businessman maximize his profit over a fixed time interval $[0, T]$? It is clear that he has to balance two effects. If he chooses $k = 0$, the firm grows rapidly but he makes no profit. If he chooses $k = 1$, he is taking everything out of the business and it cannot increase in size. What he can do once he has decided on a value of T is to choose k so as to maximize the expression

$$p(T) = \frac{kc}{(1 - k)} [e^{a(1-k)T} - 1]. \tag{15.6}$$

There are some interesting aspects to this problem which we shall discuss in the exercises below.

EXERCISES

1. For large T determine the approximate value of k which maximizes by writing

$$p(T) \cong \frac{kce^{a(1-k)T}}{(1 - k)}.$$

   Call this value $k(T)$. Is the value unique?

2. As T increases, does $k(T)$ increase or decrease?

3. For small T determine the approximate value of k by writing

$$p(T) \cong kc[aT + (aT)^2(1 - k)/2].$$

   Show that $k = 1$ until T attains a certain value.

1.16 MATHEMATICAL ECONOMICS: STEEL PRODUCTION AND BOTTLENECK PROCESSES

Let us next consider a simple model of the production of steel. Let

(16.1) $x_1(t) =$ the quantity of steel available at time t, the steel stockpile,

$x_2(t) =$ the steel plant capacity at time t.

Assuming an abundant supply of all of the materials required to produce steel, we can take the rate of production of steel to be directly proportional to steel plant capacity.

This yields the equation

$$\frac{dx_1}{dt} = a_1 x_2, \qquad x_1(0) = c_1, \tag{16.2}$$

where c_1 is the stockpile at some initial time. From this we see that we increase the rate of steel production by increasing the steel-producing capacity. But this requires steel. Let us suppose that we devote a certain fraction of the steel stockpile available at any time to the task of increasing steel capacity and that the rate of this increase is directly proportional to the quantity of steel. Then we obtain the equation

$$\frac{dx_2}{dt} = a_2 x_1, \qquad x_2(0) = c_2, \tag{16.3}$$

where c_2 is the initial plant capacity. Since we are now using steel in this fashion, Eq. (16.2) must be replaced by

$$\frac{dx_1}{dt} = a_1 x_2 - b_1 x_1, \qquad x_1(0) = c_1. \tag{16.4}$$

An interesting problem now is that of determining the quantity of steel to be allocated for increasing plant capacity which will maximize $x_1(T)$, the steel stockpile available at a specific time T.

An unrealistic feature of the foregoing is the assumption that the rate of increase in capacity is directly proportional to the rate of allocation of the steel stockpile to this effort. Since there are other requirements for the construction of steel plants, such as manpower, it is more reasonable to write

$$\frac{dx_2}{dt} = \min\,[a_2 x_1, a_3], \qquad x_2(0) = c_2. \tag{16.5}$$

Here the notation min $[a, b]$ denotes the minimum of the two nonnegative

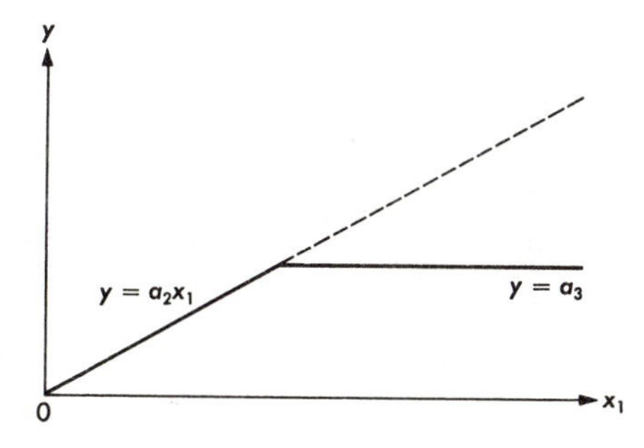

Figure 1.12

quantities a and b. As a function of x_1, min $[a_2x_1, a_3]$ has the form shown in Fig. 1.12.

This is a typical "bottleneck process," where the rate of production depends on the scarcest resource. Processes of this nature are frequently encountered in the study of industrial and economic complexes. Observe that there can be a significant deviation from proportional increase.

A number of problems concerning the most efficient production of steel can be formulated in terms of mathematical models of the foregoing type. This is a significant part of modern control theory.

1.17 DRUG KINETICS

Suppose that an organ is immersed in a large vat of a solution containing a chemical that penetrates into the organ over time. Let $u(t)$ denote the concentration of the chemical in the organ at time t and let a denote the constant concentration of the chemical in the solution. The assumption that the vat is large permits us to suppose that a remains constant.

A plausible preliminary hypothesis is that the rate of increase of $u(t)$ is directly proportional to the difference between the two concentrations, i.e.,

$$\frac{du}{dt} = k(a - u), \qquad u(0) = 0. \tag{17.1}$$

As before, we see that the solution is

$$u = a(1 - e^{-kt}), \tag{17.2}$$

under the assumption that $a - u > 0$ for $t \geq 0$. As we can see, (17.2) satisfies this condition.

Many important types of chemical reactions in the biomedical domain are covered by the foregoing model. However, many other equally important reactions are not adequately described by this simple law of proportionality. For example, we can meet equations such as

$$\frac{du}{dt} = -k, \qquad u(0) = c, \tag{17.3a}$$

$$\frac{du}{dt} = -ku^2, \qquad u(0) = c, \tag{17.3b}$$

$$\frac{du}{dt} = -\frac{ku}{(u + a)}, \qquad u(0) = c, \tag{17.3c}$$

the last of which is known as the Michaelis-Menten equation.

If two or more chemicals are present in the solution in the vat, we can have far more complex interactions. This area of drug kinetics is also a fertile source of interesting classes of differential equations.

1.18 ONE-DIMENSIONAL NEUTRON TRANSPORT THEORY

An examination of a simple version of a neutron transport process will introduce a new class of differential equations with some interesting and important properties. In this chapter we shall formulate the process in mathematical terms, but defer the solution until Chapter 2.

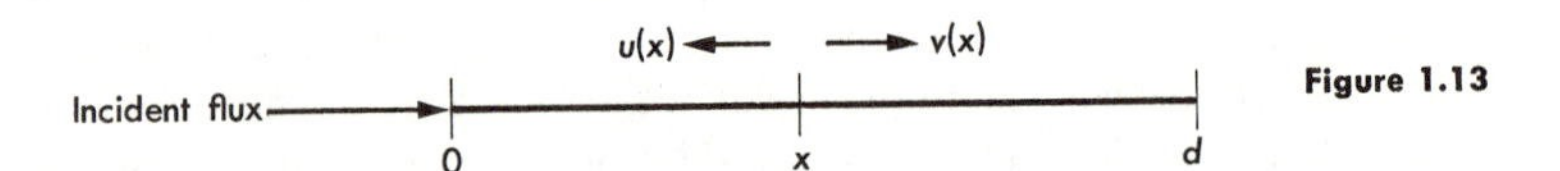

Figure 1.13

Consider a one-dimensional rod, as shown in Fig. 1.13, and suppose that particles, which we think of as neutrons, can traverse the rod in either direction. As these neutrons travel through the rod, they interact with the medium. When an interaction occurs, the original neutron disappears and is replaced by two others, one moving to the right and one to the left. This is a simple version of nuclear fission.

Assuming that there is a unit incident flux of neutrons at one end, say the left end, we wish to determine the internal fluxes to the right and to the left at a point x, and, in particular, the reflected and transmitted fluxes. We use the term "flux" here to denote the intensity of neutron flow per unit time. We must make a number of simplifying assumptions in order to treat this problem by means of differential equations. One way to proceed is the following.

We suppose that, when a neutron passes through an infinitesimal length Δ in any direction at any place in the rod, there is a probability of $1 - p\Delta$ that no interaction will take place and a probability $p\Delta$ that fission will occur. For the reader who feels shaky as far as the application of probability theory is concerned, it is sufficient to think in terms of average values. When we say that there is a probability $1 - p\Delta$ of no interaction, we mean that a fraction, $1 - p\Delta$, of the neutrons pass through the interval of length Δ with no interaction, and so on. When an interaction *does* occur, it will result in two neutrons, one moving to the right and one to the left.

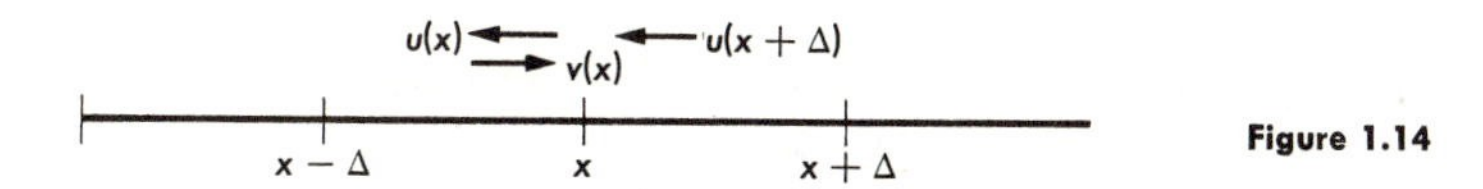

Figure 1.14

Let us assume, as indicated in Fig. 1.13, that

(18.1)
$$u(x) = \text{flux to the left at } x,$$
$$v(x) = \text{flux to the right at } x.$$

To obtain differential equations for the functions $u(x)$ and $v(x)$, we argue as follows. The flux to the left at x is a consequence of the corresponding flux to the left at $x + \Delta$ which gets through the interval $[x, x + \Delta]$ without interaction, together with the result of interaction in this interval due to this flux and the flux to the right at x which induces fission in the same portion of the rod $[x, x + \Delta]$. See Fig. 1.14. Writing this out, we have

(18.2)
$$u(x) = (1 - p\Delta)u(x + \Delta) + p\,\Delta u(x + \Delta) + p\,\Delta v(x),$$

and, similarly,

(18.3)
$$v(x) = (1 - p\Delta)v(x - \Delta) + p\,\Delta v(x - \Delta) + p\,\Delta u(x).$$

Writing

(18.4)
$$u(x + \Delta) = u(x) + \Delta u'(x),$$
$$v(x - \Delta) = v(x) - \Delta v'(x)$$

(ignoring terms in Δ^2 as usual), and simplifying, we find that Eqs. (18.2) and (18.3) yield

(18.5)
$$u'(x) = -pv(x),$$
$$v'(x) = pu(x).$$

According to our assumptions, we have an incident flux at the left end and none at the right end. Hence we have boundary conditions

(18.6) $$v(0) = 1, \qquad u(d) = 0.$$

The reflected flux is $u(0)$, and the transmitted flux is $v(d)$.

To solve a system of two simultaneous differential equations requires a bit more than the direct method used in the previous sections. Furthermore, note that the conditions are no longer initial conditions. We shall return to this interesting and important type of problem again in Chapter 2.

1.19 ONE-DIMENSIONAL NEUTRON TRANSPORT THEORY, AN ALTERNATIVE APPROACH

Suppose that, in connection with the design of the shielding of a nuclear reactor, we stated that we were not particularly interested in internal fluxes in the shield, but only in the reflected and transmitted fluxes. Could we obtain equations for these quantities directly without the intermediary of the internal fluxes?

Let us introduce as a variable y, the length of the rod, and write

(19.1) $r(y) =$ the intensity of flux reflected from a rod of length y, due to a unit incident flux.

Then, to obtain an equation for $r(y)$ we reason as follows:

(19.2) Incident flux → y $y - \Delta$ 0

The flux incident at the left-hand side traverses the infinitesimal interval $[y, y - \Delta]$ with the possibilities of either interaction or noninteraction. An interaction produces an immediate reflected flux of $p\Delta$, plus the reflected flux due to the intensity $p\Delta$ incident upon the length $y - \Delta$, namely $p\,\Delta r(y - \Delta)$. We do not have to worry about any further interactions because they affect the intensity by terms of order Δ^2, which we ignore by comparison with Δ.

The intensity incident upon $y - \Delta$ as a result of no interaction in the interval $[y, y - \Delta]$ is $1 - p\Delta$. The immediate reflected intensity is $(1 - p\Delta)r(y - \Delta)$. In traversing the interval $[y, y - \Delta]$ this flux suffers a diminution with the result that $(1 - p\Delta)^2 r(y - \Delta)$ emerges at y and engages in interaction, producing a total contribution

(19.3) $$(1 - p\Delta)r(y - \Delta)[p\Delta + p\,\Delta r(y - \Delta)].$$

The first term in this result corresponds to the neutrons going to the left as the result of fission, and the second corresponds to the reflection from $y - \Delta$ due to the neutrons going to the right.

Adding up all of the contributions, we obtain

$$r(y) = [p\Delta + p\,\Delta r(y - \Delta)] + (1 - p\Delta)^2 r(y - \Delta) + (1 - p\Delta)r(y - \Delta)[p\Delta + p\,\Delta r(y - \Delta)]. \tag{19.4}$$

Writing

$$r(y - \Delta) = r(y) - \Delta r'(y) \tag{19.5}$$

and keeping only terms, of order Δ at most, we find that after simplification Eq. (19.4) yields

$$r'(y) = p + pr(y)^2. \tag{19.6}$$

Since a rod of thickness zero produces no reflected intensity, we impose the initial condition $r(0) = 0$.

EXERCISES

1. Writing Eq. (19.6) in the form

$$\frac{r'(y)}{1 + r(y)^2} = p$$

and integrating both sides with respect to y, show that $r(y) = \tan py$.

2. Hence show that there is a "critical length," $y = \pi/2p$, which yields an infinite reflected flux. This simple calculation shows the feasibility of a nuclear device which can be used either constructively or destructively.

1.20 DISCUSSION

Now that we have shown by means of the foregoing examples that differential equations are a natural and versatile tool to use in the description of physical phenomena, the question arises as to the next step to take. There are several ways in which we can proceed. We can, first of all, attempt to find classes of differential equations which can be resolved in reasonably simple fashion. Secondly, we can ask when various classes of differential equations have solutions and, if so, how many. Thirdly, we can seek to determine algorithms which will enable us to obtain the numerical solution even when explicit analytic solutions do not exist.

In the next chapter, we shall study the most important class of differential equations, as far as both theory and application are concerned, linear differential equations with constant coefficients. Examples were met in Sections 1.5, 1.6, 1.7, 1.13, and 1.18.

1.21 DIRECT USE OF DIFFERENTIAL EQUATIONS

In the previous sections we elaborated on the theme that mathematics is a language which can be used to express scientific ideas in a convenient and often useful fashion. Once a problem has been translated into a differential equation, we can proceed to solve the equation either analytically or computationally and then use the solution to resolve the original problem.

This is occasionally a difficult task and almost always time-consuming and energy-consuming. It is therefore important to point out that in many cases a good deal of useful information can be obtained easily from the differential equation using only elementary ideas of calculus. These preliminary results can be used to test the validity of the mathematical formulation and the numerical calculations. Let us proceed to give some examples.

The equation

$$\frac{du}{dt} + au = 1, \qquad u(0) = c, \tag{21.1}$$

where $a > 0$, can arise in connection with studies of penetration of a particle into a medium. One way of determining the maximum depth of penetration is to find u as a function of t, as we do in the following chapter, and use the explicit expression to determine this limiting behavior. But it is quicker and easier to read the result from Eq. (21.1). At this stage the reasoning is purely formal. Subsequent developments which will be presented in Chapter 2 will make it rigorous.

As u approaches its maximum value, it is plausible that du/dt approaches zero. Hence from Eq. (21.1) we see that

$$u_{\max} = 1/a. \tag{21.2}$$

This value, as we shall see, corresponds to $t = \infty$ and is therefore not attained at any finite time.

Consider next the equation

$$\frac{du}{dt} = au - bu^2, \qquad u(0) = c, \tag{21.3}$$

where $a, b > 0$. We discussed this equation in some detail above. If $u(t)$

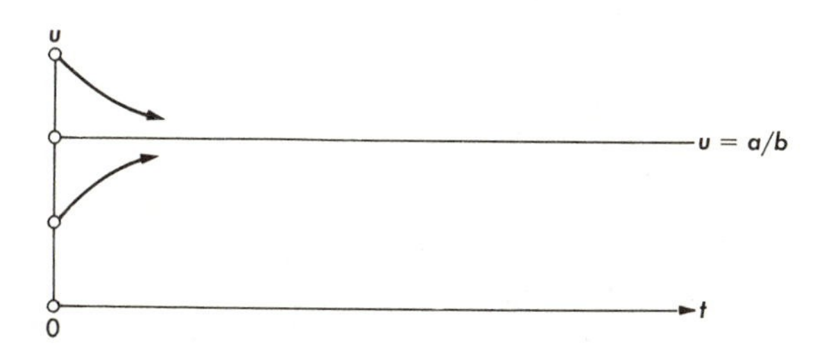

Figure 1.15

represents the population at time t, then a first question of interest is that of the steady state. For the population to remain constant, we must have $du/dt = 0$. Hence from (21.3) we see that steady-state populations are either a/b or the uninteresting value of zero.

Suppose that we are interested in knowing when the population is increasing. If c is initially less than a/b, we see that

$$au - bu^2 = u(a - bu)$$

is positive. Hence $du/dt > 0$, and u is monotone increasing, as illustrated in Fig. 1.15. If c is initially greater than a/b, then du/dt is negative and u is initially decreasing.

Can the solution $u(t)$ cross the line $u = a/b$ in either case? The answer is easily seen to be negative. At the point P we have $u = a/b$ (Fig. 1.16), which, according to the differential equation, makes $du/dt = 0$.

As we shall see subsequently, the uniqueness theorem for solutions of Eq. (21.3) immediately tells us that this crossing is impossible.

As a third example of these simple techniques consider the equation

$$u'' + u = 0, \tag{21.4}$$

associated with the motion of a linear spring or a pendulum with small displacement. We expect a pendulum to swing from one extreme position to the other in a smooth fashion. We would be very surprised to find that the pendulum swung part way down, then returned part way up, then went

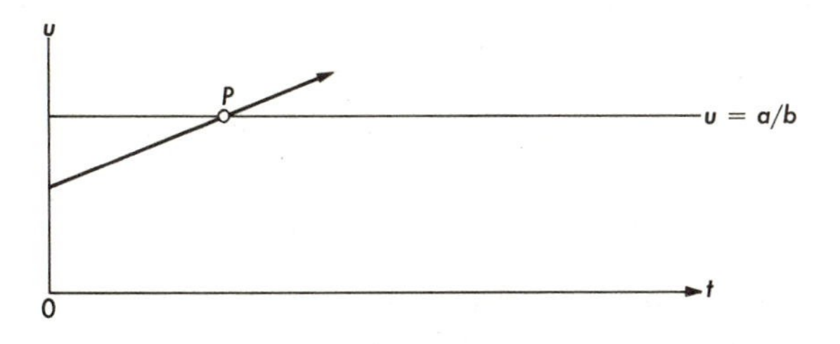

Figure 1.16

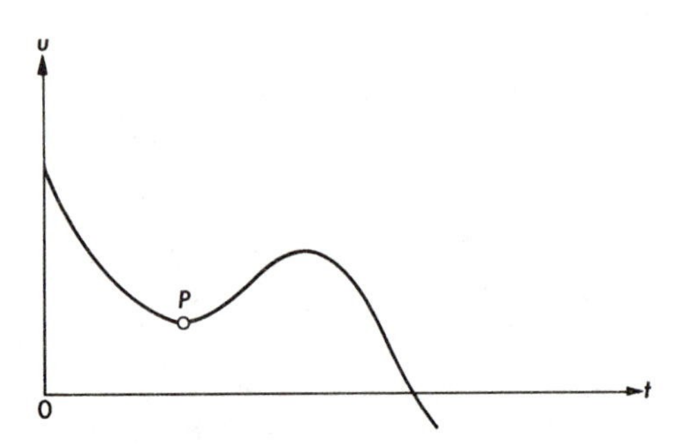

Figure 1.17

through the rest position. A time history of such a behavior would have the form shown in Fig. 1.17.

From the differential equation we can see immediately that the point P cannot exist. For at P, u has a relative minimum, which means that u'' is positive there. But from the differential equation (21.4) we see that at P,

$$u'' = -u < 0. \tag{21.5}$$

Finally, we know that in the motion of a pendulum energy is conserved with a constant interchange between potential and kinetic energy. Can we see that directly from (21.4)? Indeed we can. The total energy is

$$E = u'^2 + u^2, \tag{21.6}$$

in the units used in Eq. (21.4).

We see that

$$\frac{dE}{dt} = 2u'u'' + 2uu' = 2u'(u'' + u) = 0. \tag{21.7}$$

Hence there is conservation of energy: E is a constant. From this we see that as the potential energy increases, the kinetic energy decreases.

EXERCISES

---

1. If $u'' + au' + bu = 1$, what are the possible steady-state values of u?
2. Given that

$$u' = au - bu^2, \qquad a, b > 0,$$

show that $u(t)$ has a point of inflection at a t-value where $u = a/2b$. Does it have more than one point of inflection?

3. Consider the equation

$$u' - u = t, \qquad u(0) = 1.$$

Show that if a solution intersects the line $u = t$, it intersects it at a point where u has a relative minimum.

4. Consider the equation $u'' - u = 0$. Show that no nontrivial solution can assume the value zero more than once.

5. Establish the same result for

$$u'' - g(t)u = 0$$

under the assumption that $g(t)$ is uniformly positive for $t \geq 0$.

6. Establish the same result for

$$u'' - (u + u^3) = 0$$

and for

$$u'' - t^m u^n = 0, \qquad m, n > 0.$$

7. Consider the equation

$$u'' + \sin u = 0.$$

Given that $\pi > u(0) > 0$, $u'(0) < 0$, show that u steadily decreases until it reaches zero.

8. Show that energy is conserved in the system described by the equation of Exercise 7.

1.22 PARTICULAR SOLUTIONS

Continuing the theme of the previous section, in many cases we are more interested in particular solutions of differential equations than in general solutions. In other words, we may be more concerned with a particular behavior of a physical system than with all possible behaviors.

Let us discuss some examples of this. A system with an external periodic influence can often be described by an equation of the form

$$u'' + u = \sin bt. \tag{22.1}$$

Does the external vibration set up an internal vibration with the same frequency? Can we have a solution of Eq. (22.1) of the form $u = a \sin bt$.

Substituting in (22.1), we have

$$-ab^2 \sin bt + a \sin bt = \sin bt, \tag{22.2}$$

whence

$$a = \frac{1}{1 - b^2}. \tag{22.3}$$

Therefore, if $b \neq 1$ the answer is affirmative. What happens when $b = 1$ will be discussed in the next chapter.

Similarly, given the equation

$$u'' - t^m u^n = 0, \tag{22.4}$$

an equation arising in both astrophysics and quantum mechanics, it is of interest to know whether or not a solution of the form at^b exists. Substituting in Eq. (22.4), we have

$$ab(b - 1)t^{b-2} - t^m(a^n t^{bn}) = 0, \tag{22.5}$$

whence the conditions which determine a and b are

$$\begin{aligned} ab(b - 1) &= a^n, \\ b - 2 &= m + bn. \end{aligned} \tag{22.6}$$

If $n \neq 1$, there is a unique solution.

EXERCISES

1. Does $u'' + ku' + u = \sin bt$ have a solution of the form $a \sin bt$ for all b if $k \neq 0$?
2. Does $u'' + au' + bu = 0$ always possess a solution of the form $e^{\lambda t}$, where λ may be complex?
3. Show that at most two such values of λ exist in Exercise 2. When is there only one?
4. Given that $e^{\lambda_1 t}$ and $e^{\lambda_2 t}$ are particular solutions, show that $b_1 e^{\lambda_1 t} + b_2 e^{\lambda_2 t}$ are particular solutions for any constants b_1 and b_2.
5. Can b_1 and b_2 in Exercise 4 be chosen to satisfy any two conditions of the form $u(0) = c_1$, $u'(0) = c_2$, provided that $\lambda_1 \neq \lambda_2$?
6. Find a linear differential equation satisfied by

$$u = c_1 \cos t + c_2 \sin t$$

in the following fashion. We have

$$u' = -c_1 \sin t + c_2 \cos t,$$
$$u'' = -c_1 \cos t - c_2 \sin t.$$

Eliminating 1, c_1 and c_2 from these three simultaneous equations, we have the determinantal relation

$$\begin{vmatrix} u & \cos t & \sin t \\ u' & -\sin t & \cos t \\ u'' & -\cos t & -\sin t \end{vmatrix} = 0.$$

Show that this equation reduces to $u'' + u = 0$. Is there an easier way of obtaining this equation?

BIBLIOGRAPHY AND COMMENTS

Section 1.5. The translation of Newton's laws of motion into differential equations provided the stimulus for a burst of scientific and mathematical activity in the eighteenth and nineteenth centuries. The purely analytic research centering about celestial mechanics was the antecedent to the greater part of contemporary research in the seemingly far-removed domains of algebra and topology.

Section 1.6. In Section 1.5 we employed the traditional imbedding technique in order to answer the original question concerning the maximum height attained by the ball. In Section 1.6 we used a different technique of imbedding, called "invariant imbedding," a method which we used again in Section 1.19. The point we wish to stress is that it is to be expected that there will be various alternative formulations of the same physical process, each with its own advantages and disadvantages. The well-trained mathematician employs them all at appropriate times.

Section 1.7. The mathematical study of the growth of a population and, in particular, of the interactions of two hostile or symbiotic populations contains many intriguing features. Its importance in a world beginning to suffer from an over-population and undersupply of food is apparent. For some early and still significant investigations, see:

LOTKA, A. J., *Elements of Physical Biology*, Williams and Wilkins, Baltimore, 1925.

VOLTERRA, V., *Leçons sur la théorie mathématique de la lutte pour la vie*, Gauthiers-Villars, Paris, 1931.

For more sophisticated treatments on a higher analytic level, see:

BHARUCHA-REID, A. T., *Elements of the Theory of Markov Processes and Their Applications*, McGraw-Hill, New York, 1960.

BAILEY, N. T. J., *The Elements of Stochastic Processes with Applications to the Natural Sciences*, John Wiley, New York, 1964.
Many other references of interest will be found in these volumes.

Section 1.8. The study of inverse problems is basic to all parts of science. In particular, its domain includes the investigation of the identification of systems. An idea of the kinds of analytic questions and the types of results that may be obtained may be gained by perusing the following:

BELLMAN, R., and R. KALABA, *Quasilinearization and Nonlinear Boundary Value Problems*, American Elsevier, New York, 1965.
Systematic investigations of inverse problems is now possible because of the availability of digital computers.

Section 1.9. For an introductory account of stability theory, see:

BELLMAN, R., *Stability Theory of Differential Equations*, Dover Publications, New York, 1969.

Section 1.14. The reader interested in further details concerning analog computers may wish to consult:

TECHNICAL EDUCATION AND MANAGEMENT, INC., *Computer Basics, Volume One: Introduction to Analog Computers*, Howard S. Sams, Indianapolis, Ind., 1962.

Section 1.16. For a discussion of bottleneck processes, see:

BELLMAN, R., *Dynamic Programming*, Princeton University Press, Princeton, N.J., 1957.

Section 1.17. For a detailed account of mathematical models in this area, see:

RESCIGNO, A., and G. SEGRE, *Drug and Tracer Kinetics*, Blaisdell, Waltham, Mass., 1966.

GRODINS, F. S., *Control Theory and Biological Systems*, Columbia University Press, New York, 1963.

Section 1.18. For an introduction to the classical treatment of neutron transport theory, see:

WING, G. M., *An Introduction to Transport Theory*, Wiley, New York, 1962.
The important point in this section is that many classes of physical problems lead to conditions on the solution which are not of the initial-value type.

Section 1.19. Here the theory of invariant imbedding is employed again to produce a differential equation with an initial condition. The reason for our intense interest in initial-value problems will be discussed in Chapter 4.

CHAPTER 2

Linear Differential Equations and their Solutions in Compact Forms

2.1 INTRODUCTION

In this chapter we wish to study in careful detail a particularly simple, yet extremely important, type of differential equation, the equation

$$y'' + ay' + by = g(x), \tag{1.1}$$

where the coefficients a and b are constants. A complete analysis of the behavior of the solution subject to various types of initial and boundary conditions is possible as a consequence of the fortunate fact that an explicit representation of the solution where $g(x)$ is equal to zero can be obtained in terms of exponential functions.

The significance of Eq. (1.1) resides in its constant appearance in all parts of pure and applied mathematics and in its frequent use for approximation purposes. With its aid we can illustrate a variety of important and interesting mathematical and physical phenomena. The student who has mastered all aspects of the behavior of the solution as a function both of the independent variable and the coefficients in the equation has a good hold on mathematical physics.

As the reader will see upon turning to Section 2.7, there is little trouble involved in finding *a* solution for the equation

$$y'' + ay' + by = 0, \tag{1.2}$$

where y is subject to the initial conditions

$$y(0) = c_1, \qquad y'(0) = c_2. \tag{1.3}$$

The question that arises is whether it is *the* solution. If Eq. (1.2) is to be used to describe a physical system, we want to be sure that the solution has physical significance. A second reason for adopting a general approach lies in the fact that the method used in Section 2.7 does not carry over to the more general equation (1.1). Consequently, from the standpoint of both understanding and expediency, it is important to study a more powerful method for handling Eq. (1.2). This more flexible approach will simultaneously establish the fact that there is exactly one solution for Eq. (1.2) subject to the conditions (1.3) and yield the general solution for Eq. (1.1).

At the close of this chapter we will study the question of obtaining a solution for Eq. (1.2) subject to the conditions

$$y(0) = c_1, \qquad y(T) = c_2. \tag{1.4}$$

2.2 A SIMPLE UNIQUENESS THEOREM

Let us begin with the simplest differential equation,

$$\frac{dy}{dx} = g(x). \tag{2.1}$$

If we assume that $g(x)$ is continuous for $x \geq 0$, we know that

$$y = \int_0^x g(x_1)\,dx_1 \tag{2.2}$$

is a solution for $x \geq 0$. If we impose the initial condition $y(0) = c$, we have a solution of Eq. (2.1):

$$y = \int_0^x g(x_1)\,dx_1 + c. \tag{2.3}$$

Is this the only solution satisfying the condition $y(0) = c$?

Suppose that y_1 were another solution of (2.1). Then, by assumption,

$$\frac{dy_1}{dx} = g(x). \tag{2.4}$$

Using Eqs. (2.1) and (2.4), we have

$$\frac{dy}{dx} - \frac{dy_1}{dx} = 0, \qquad \frac{d}{dx}(y - y_1) = 0. \tag{2.5}$$

This asserts that the derivative of the function $y - y_1$ is zero for $x \geq 0$. We thus conclude that $y - y_1$ is a constant for $x \geq 0$. Perhaps the simplest proof of this basic fact uses the mean-value theorem of calculus. Therefore, we know that (2.3) is the general solution of (2.1), with c an arbitrary constant. By this we mean that *every* solution of Eq. (2.1) is a function of the form given in (2.3). If we now use the fact that y is known at $x = 0$, then c is determined, and is equal to $y(0)$.

We have established a *uniqueness* theorem. The differential equation (2.1) plus an initial condition possesses only one solution. The fact that it possesses at least one solution is an *existence* theorem.

An essential step in our proof was the derivation of the second equation in (2.5) from the first. We used the fact that the derivative is a linear operation, i.e.,

$$\frac{d}{dx}(y_1 + y_2) = \frac{dy_1}{dx} + \frac{dy_2}{dx}. \tag{2.6}$$

We shall use this important property repeatedly below, and it is indeed the key to the solution of equations of the type appearing in (1.1).

2.3 THE EQUATION $y' + ay = g(x)$

Let us now turn to the equation

$$y' + ay = g(x), \qquad y(0) = c. \tag{3.1}$$

We call this a first-order equation because it contains the unknown function and its first derivative, and no higher derivatives. It is called a linear first-order equation because y and y' appear in linear fashion. To solve Eq. (3.1), we introduce a simple device, based on the special property of the exponential with respect to differentiation. We recall that for any differentiable function y and any constant a, we have

$$\frac{d}{dx}(e^{ax}y) = e^{ax}y' + ae^{ax}y. \tag{3.2}$$

Hence, if we multiply (3.1) through by e^{ax}, then the equation takes the form

$$e^{ax}(y' + ay) = e^{ax}g(x), \qquad y(0) = c, \tag{3.3}$$

$$\frac{d}{dx}(e^{ax}y) = e^{ax}g(x).$$

Upon integrating between 0 and x, we obtain the relation

$$e^{ax}y - c = \int_0^x e^{ax_1}g(x_1)\,dx_1 \tag{3.4}$$

(since at $x = 0$, $e^{ax}y$ assumes the value c). Finally, we obtain the explicit solution for (3.1):

$$y = e^{-ax} \int_0^x e^{ax_1} g(x_1)\, dx_1 + ce^{-ax}, \tag{3.5}$$

or

$$y = \int_0^x e^{-a(x-x_1)} g(x_1)\, dx + ce^{-ax}. \tag{3.6}$$

The direct approach sketched above simultaneously establishes the existence and uniqueness of the solution of Eq. (3.1).

EXERCISES

1. Find the solution of

   a) $y' + 2y = x, \quad y(0) = 1,$
   b) $y' - y = \sin x, \quad y(0) = 2.$

2. Find the general solution of

   a) $y' + 2y = x^2,$
   b) $y' - y = \sin 2x.$

3. Show that the solution of

   $$y' + ay = b, \qquad y(0) = c,$$

   is given by

   $$y = \left(c - \frac{b}{a}\right) e^{-ax} + \frac{b}{a}.$$

4. Obtain the general solution of

   $$y' + ay = e^{bx}$$

   for $b \neq -a$, and then for $b = -a$.

5. Show that the solution of

   $$y' + ay = g(x), \qquad y(0) = c,$$

   may be written in the form $y = y_1 + y_2$, where

   $$y_1' + ay_1 = g(x), \qquad y_1(0) = 0,$$
   $$y_2' + ay_2 = 0, \qquad y_2(0) = c.$$

6. Show that the solution of

$$y' + ay = 0,$$

with the condition

$$\int_0^b y\, dx_1 = c,$$

is given by

$$y = \begin{cases} \dfrac{cae^{-ax}}{(1 - e^{-ab})}, & a \neq 0, \\ \dfrac{c}{b}, & a = 0. \end{cases}$$

[*Hint:* Begin with the equation $y' + ay = 0$, $y(0) = d$, where d is an unknown.]

7. How many solutions are there of $y' + ay = b$ satisfying the condition that

$$\lim_{x \to \infty} y = b/a.$$

Consider first the case $a > 0$ and then $a < 0$.

8. Obtain the solution of

$$y' + ay = \int_0^b y\, dx_1, \qquad y(0) = c.$$

[*Hint:* First set $\int_0^b y\, dx_1 = m$. Solve for y in terms of m and then obtain an equation for m.]

9. Similarly, solve

$$y' + ay = \int_0^1 x_1 y\, dx_1, \qquad y(0) = 1,$$

$$y' + ay = \int_0^1 f(x_1) y\, dx_1, \qquad y(0) = 1.$$

10. Show that $y' + ay = 1$ has a solution of the form $y = k$, a constant, and obtain the value of k.

11. Show that $y' + ay = x$ has a solution of the form $y = k_1 + k_2 x$, where k_1 and k_2 are constants, and obtain the values of k_1 and k_2. [*Hint:* Obtain the result using (3.5) directly, without using this formula.]

12. Show that $y' + ay = p(x)$, where $p(x)$ is a polynomial of degree n in x, always has a solution which is a polynomial of degree n, provided that $a \neq 0$. What happens if $a = 0$?

13. Solve the integral equation

$$u(x) + \int_0^x u(x_1)\,dx_1 = g(x)$$

by reducing it to a differential equation. [*Hint:* Let $v = \int_0^x u(x_1)\,dx_1$.]

14. The equation $y' + ay = \sin \omega x$ has a solution of the form $b_1 \sin \omega x + b_2 \cos \omega x$.

15. If $a > 0$, the solution of $y' + ay = \sin \omega x$, $y(0) = c$, tends to this solution as $x \to \infty$ regardless of the value of c.

16. Consider the particular solution of

$$y' + ay = e^{bx}, \qquad y = \frac{e^{bx}}{b + a}.$$

Show how a particular solution can be obtained for

$$y' + ay = x$$

by differentiation with respect to b, followed by setting $b = 0$.

17. Let y_1 be the solution of

$$y' + ay = g(x), \qquad y(0) = c_1,$$

and y_2 the solution with $y(0) = c_2$. If $c_1 \neq c_2$, can $y_1(x) = y_2(x)$ for any $x > 0$?

2.4 THE EQUATION $y' + a(x)y = g(x)$

Let us next consider the equation

$$y' + a(x)y = g(x), \qquad y(0) = c, \tag{4.1}$$

where the coefficient of y is a function of x. Fortunately, this more complex equation can be handled by means of the same simple device used in Section 2.3, and we can once again obtain the general solution. We recall that

$$\begin{aligned} \frac{d}{dx}\left(e^{\int_0^x a(x_1)dx_1} y\right) &= e^{\int_0^x a(x_1)dx_1} y' + e^{\int_0^x a(x_1)dx_1} a(x)y \\ &= e^{\int_0^x a(x_1)dx_1} [y' + a(x)y]. \end{aligned} \tag{4.2}$$

Hence, if we multiply (4.1) through by the term $e^{\int_0^x a(x_1)dx_1}$, the equation takes the form

$$(4.3) \qquad e^{\int_0^x a(x_1)dx_1}(y' + a(x)y) = e^{\int_0^x a(x_1)dx_1}g(x),$$
$$\frac{d}{dx}\left(e^{\int_0^x a(x_1)dx_1}y\right) = e^{\int_0^x a(x_1)dx_1}g(x).$$

Integrating between 0 and x, we obtain the relation

$$(4.4) \qquad e^{\int_0^x a(x_1)dx_1}y - c = \int_0^x \left[e^{\int_0^{x_2} a(x_1)dx_1}g(x_2)\right] dx_2,$$

taking account of the fact that $y = c$ at $x = 0$.

Solving for y, we obtain the important result

$$(4.5) \qquad y = ce^{-\int_0^x a(x_1)dx_1} + e^{-\int_0^x a(x_1)dx_1}\int_0^x e^{\int_0^{x_2} a(x_1)dx_1}g(x_2)\, dx_2.$$

If we wish we can write this in the form

$$(4.6) \qquad y = ce^{-\int_0^x a(x_1)dx_1} + \int_0^x e^{-\int_{x_2}^x a(x_1)dx_1}g(x_2)\, dx_2.$$

EXERCISES

1. Obtain the solution of $y' + a(x)y = g(x)$, $y(b) = c$.
2. Show that the solution of $y' + xy = 1$, $y(0) = 0$, is given by
$$e^{-x^2/2}\int_0^x e^{x_1^2/2}\, dx_1.$$
3. Show that $y' + xy = x^2 - x + 1$ possesses a solution which is a quadratic in x.
4. Obtain the solution of $y' + a(x)y = 0$ subject to the condition
$$\int_0^b y\, dx_1 = c.$$
5. Obtain the solution of $y' + a(x) = \int_0^b y\, dx_1$, $y(0) = c$.
6. Solve the integral equation $u(x) + \int_0^x f(x_1)u(x_1)\, dx_1 = g(x)$. [*Hint:* Let $v = \int_0^x f(x_1)u(x_1)\, dx_1$, and multiply the equation through by $f(x)$.]

7. Solve the integral equation

$$u(x) + \int_0^x e^{a(x-x_1)} u(x_1)\, dx_1 = g(x)$$

by converting it into a differential equation.

8. Let y_1 be the solution of

$$\frac{dy_1}{dx} + a(x)y_1 = g_1(x), \qquad y_1(0) = c_1$$

and y_2 be the solution of

$$\frac{dy_2}{dx} + a(x)y_1 = g_2(x), \qquad y_2(0) = c_2.$$

If $c_1 > c_2$ and $g_1 = g_2$, then $y_1 > y_2$ for $x \geq 0$. [*Hint:* Use the explicit form of the solution and the positivity of the exponential function.]

9. If $c_1 = c_2$ and $g_1 > g_2$, then $y_1 > y_2$ for $x \geq 0$.

10. If $c_1 > c_2$, $g_1 > g_2$, then $y_1 > y_2$ for $x \geq 0$.

11. Let y be the solution of $dy/dx + a(x)y = g(x)$, $y(0) = c$, and z a function such that $dz/dx + a(x)z \leq g(x)$, with $z(0) = c$. Show that $z(x) \leq y(x)$ for $x \geq 0$.

12. Show that $u(t) \leq c + \int_0^t f(t_1)u(t_1)\, dt_1$ for $t \geq 0$, with c, $u(t)$ and $f(t)$ nonnegative, implies that $u(t) \leq c \exp\left(\int_0^t f(t_1)\, dt_1\right)$. [A straightforward way to establish the foregoing result is to set $v = \int_0^t f(t_1)u(t_1)\, dt_1$ and multiply the inequality through by $f(t)$ to obtain

$$f(t)u(t) \leq cf(t) + f(t)\int_0^t f(t_1)u(t_1)\, dt_1,$$

or

$$\frac{dv}{dt} \leq cf(t) + f(t)v, \qquad v(0) = 0.$$

Comparing the solution of the inequality with the solution of the equality, and using the original inequality, we obtain the stated result. In view of the importance of this inequality, sometimes called the "fundamental inequality," let us present a shorter proof. From the original inequality we have

$$\frac{u(t)}{c + \int_0^t f(t_1)u(t_1)\, dt_1} \leq 1, \qquad \frac{f(t)u(t)}{c + \int_0^t f(t_1)u(t_1)\, dt_1} \leq f(t).$$

Integrating between 0 and t, we have

$$\log\left(c + \int_0^t f(t_1)u(t_1)\,dt_1\right)\Big]_0^t \le \int_0^t f(t_1)\,dt_1.$$

Hence

$$\log\left(c + \int_0^t f(t_1)u(t_1)\,dt_1\right) - \log c \le \int_0^t f(t_1)\,dt_1,$$

or

$$c + \int_0^t f(t_1)u(t_1)\,dt_1 \le c\exp\left(\int_0^t f(t_1)\,dt_1\right)\cdot$$

Referring to the original inequality this yields the desired result

$$u(t) \le c\exp\left(\int_0^t f(t_1)\,dt_1\right).]$$

2.5 DECOUPLING OF INITIAL CONDITION AND FORCING TERM

In many physical systems a term such as $g(x)$ in (4.1) represents an external force. We shall occasionally call it a "forcing term." Examining (4.6), we see that the solution can be written as a sum of two terms, a contribution due to the initial condition, (the initial "state" of the system), and a contribution due to the forcing term. This is a characteristic of linear equations.

Specifically, the solution of (4.1) may be written in the form

$$y = y_1 + y_2, \tag{5.1}$$

where

$$y_1' + a(x)y_1 = 0, \qquad y_1(0) = c, \tag{5.2}$$

and

$$y_2' + a(x)y_2 = g(x), \qquad y_2(0) = 0. \tag{5.3}$$

2.6 DISCONTINUOUS FUNCTIONS

We have considered in the previous sections the equation

$$y' + a(x)y = g(x) \tag{6.1}$$

and given a number of examples of the form of the solution for the cases

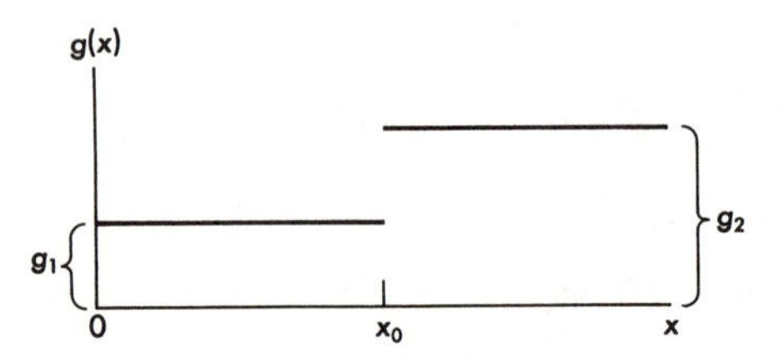

Figure 2.1

where $a(x)$ and $g(x)$ were constants, polynomials, exponentials, and trigonometric functions. These are all examples of continuous functions. In many important scientific situations we meet discontinuous functions. Let us consider two examples of this, one in which the "forcing function" $g(x)$ is discontinuous and the other in which the coefficient $a(x)$ has this property. In both cases, we shall consider the same common type of discontinuity.

Example 1. Consider the equation

$$y' + ay = g(x), \qquad y(0) = c, \tag{6.2}$$

where $g(x)$ has the form in Fig. 2.1.

Thus

$$\begin{aligned} g(x) &= g_1, \qquad 0 \le x \le x_0 \\ &= g_2, \qquad x_0 < x < \infty. \end{aligned} \tag{6.3}$$

The discontinuity is a simple jump at $x = x_0$. The function is Riemann-integrable and hence the method of procedure of Section 2.4 is valid. Let us take $c = 0$ to simplify. We know, from Section 2.5, that we can consider the solution as a sum of the solutions for $c = 0$ and $g(x) = 0$, respectively. We have, with $c = 0$,

$$y = e^{-ax} \int_0^x e^{ax_1} g(x_1)\, dx_1. \tag{6.4}$$

Consider first the case where $0 \le x \le x_0$. Then from (6.3)

$$\begin{aligned} y &= e^{-ax} \int_0^x e^{ax_1} g_1\, dx_1 = g_1 e^{-ax} \int_0^x e^{ax_1}\, dx_1 \\ &= g_1 e^{-ax}(e^{ax} - 1)/a \\ &= g_1(1 - e^{-ax})/a, \end{aligned} \tag{6.5}$$

provided that $a \neq 0$. If $a = 0$, we have directly from (6.1)

$$y = \int_0^x g(x_1)\, dx_1 = g_1 \int_0^x dx_1 = g_1 x. \tag{6.6}$$

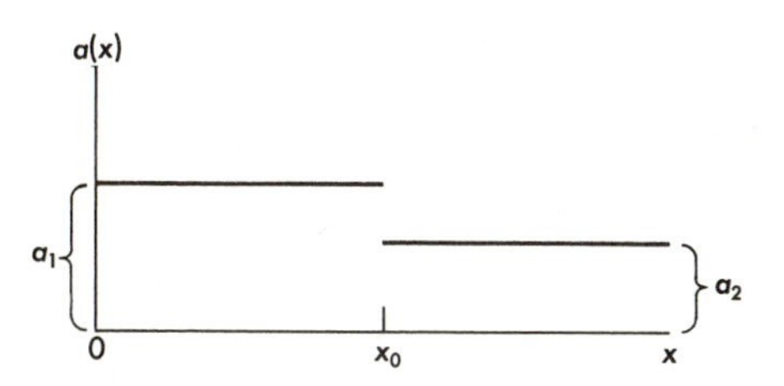

Figure 2.2

Consider next the case $x > x_0$. Write

$$
\begin{aligned}
(6.7) \qquad y &= e^{-ax} \int_0^{x_0} e^{ax_1} g(x_1)\, dx_1 + e^{-ax} \int_{x_0}^{x} e^{ax_1} g(x_1)\, dx_1 \\
&= g_1 e^{-ax} \int_0^{x_0} e^{ax_1}\, dx_1 + g_2 e^{-ax} \int_{x_0}^{x} e^{ax_1}\, dx_1 \\
&= g_1 e^{-ax}(e^{ax_0} - 1)/a + g_2 e^{-ax}(e^{ax} - e^{ax_0})/a.
\end{aligned}
$$

Observe, comparing (6.5) and (6.7), that y is continuous at $x = x_0$, despite the fact that $g(x)$ is not. The function y, however, does not possess a derivative at $x = x_0$. The differential equation then is satisfied for $0 \le x < x_0$, and for $x > x_0$. It does *not* hold at $x = x_0$. There is no solution which possesses a derivative for all $x \ge 0$.

Example 2. Consider the equation

$$(6.8) \qquad y' + a(x)y = 0, \qquad y(0) = c,$$

where

$$
\begin{aligned}
(6.9) \qquad a(x) &= a_1, \qquad 0 \le x \le x_0, \\
&= a_2, \qquad x_0 < x,
\end{aligned}
$$

as shown in Fig. 2.2.

Then

$$(6.10) \qquad y = ce^{-\int_0^x a(x_1)dx_1}$$

is a solution except at $x = x_0$. It is continuous at this point, but does not possess a derivative there. The function $\int_0^x a(x_1)\, dx_1$ has the form in Fig. 2.3.

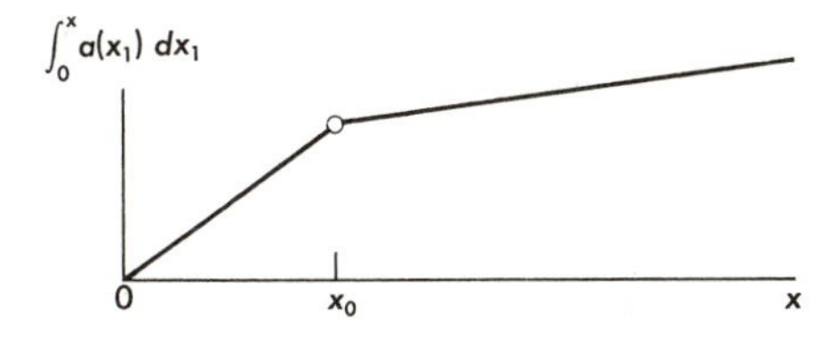

Figure 2.3

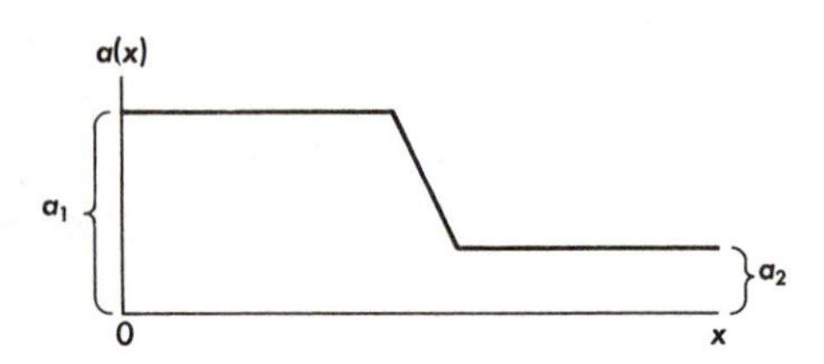

Figure 2.4

A discontinuity in the forcing term of the foregoing type corresponds to an abrupt change in the nature of the external force at $x = x_0$. A discontinuity in the coefficient corresponds to an abrupt change in the nature of the system as we go from one region to another. It turns out that it is usually very much simpler to portray these abrupt transitions by means of discontinuous functions than by means of rapidly varying functions such as, for example, Fig. 2.4.

EXERCISES

1. Find the solution of $y' + a(x)y = 0$, $y(0) = c$, for each of the following functions a.

a) $a(x) = 2, \quad 0 \le x \le 1,$
$ = 1, \quad 1 < x < \infty.$

b) $a(x) = 1, \quad 0 \le x \le 1,$
$ = 0, \quad 1 < x \le 2,$
$ = 2, \quad 2 < x \le 3,$
$ = 0, \quad 3 < x < \infty.$

c) $a(x) = x, \quad 0 \le x \le 1,$
$ = 2 - x, \quad 1 < x \le 2,$
$ = 0, \quad 2 < x < \infty.$

2. Find the solution of $y' + 2y = g(x)$, $y(0) = c$, for each of the following functions g.

a) $g(x) = 2, \quad 0 \le x \le 1,$
$ = 1, \quad 1 < x < \infty.$

b) $g(x) = \sin x, \quad 0 \le x \le \pi,$
$ = 0, \quad \pi < x < \infty.$

c) $g(x) = 1, \quad 0 \le x \le 1,$
$ = 0, \quad 1 < x \le 2,$
$ = 2, \quad 2 < x \le 3,$
$ = 0, \quad 3 < x < \infty.$

d) $$\begin{aligned} g(x) &= x, & 0 &\le x \le 1, \\ &= 1, & 1 &< x \le 2, \\ &= 3 - x, & 2 &< x \le 3, \\ &= 0, & 3 &< x < \infty. \end{aligned}$$

2.7 THE EQUATION $y'' + ay' + by = 0$

In order to treat the equation

$$y'' + ay' + by = 0 \tag{7.1}$$

in as detailed a fashion as required, we regret to say that we are forced to introduce another stratagem. There are several which may be employed. Which ones to present and in what order is largely a matter of taste. Let us begin with a method which will be useful subsequently.

Our aim is to reduce the solution of this new equation to that of a succession of equations of the type we have already studied. This is, after all, the way we deal with the special case

$$y'' = 0. \tag{7.2}$$

We integrate once, and then we integrate again, which is to say that we think of the equation in the form $(y')' = 0$. An equation of the type given in (7.1) we call a second-order linear differential equation for obvious reasons.

We proceed in the following fashion. Do there exist two constants p and q such that

$$y'' + ay' + by = \frac{d}{dx}(y' - py) - q(y' - py) \tag{7.3}$$

for any function y? Carrying out the indicated differentiation, we reduce the right-hand side to

$$y'' - py' - qy' + qpy. \tag{7.4}$$

Hence, if Eq. (7.3) is to be an identity, we must have

$$p + q = -a, \qquad pq = b. \tag{7.5}$$

These two relations are equivalent to the statement that p and q are the two roots of the quadratic equation

$$r^2 + ar + b = 0. \tag{7.6}$$

Comparing the algebraic equation (7.6) with the differential equation

(7.7) $$y'' + ay' + by = 0,$$

it is easy to see where the quadratic equation arises. This equation, (7.6), which we shall encounter again, is called the *characteristic equation* of (7.7), and its roots are called the *characteristic roots.* Let us denote these roots by r_1 and r_2, where

(7.8) $$r_1 = \frac{-a + \sqrt{a^2 - 4b}}{2}, \qquad r_2 = \frac{-a - \sqrt{a^2 - 4b}}{2}.$$

If $a^2 \neq 4b$, the roots are unequal. Let us assume for the moment that this is the case.

Returning to Eq. (7.1), let us write it in the form

(7.9) $$\frac{d}{dx}(y' - r_1 y) - r_2(y' - r_1 y) = 0.$$

This in turn can be written as

(7.10) $$\frac{dw}{dx} - r_2 w = 0,$$

where we have introduced the new variable $w = y' - r_1 y$. We obtained in Chapter 1 the general solution of Eq. (7.10), namely

(7.11) $$w = ce^{r_2 x},$$

where c is an arbitrary constant, the value of $w(0)$.

To derive from (7.11) the desired function y, we return to the fact that $w = y' - r_1 y$. Replacing w by its value in (7.11) yields the equation

(7.12) $$y' - r_1 y = ce^{r_2 x}.$$

As we know from Section 2.3, the solution of this first-order linear equation is given by

(7.13) $$y = de^{r_1 x} + e^{r_1 x} \int_0^x e^{-r_1 x_1}(ce^{r_2 x_1})\, dx_1,$$

where d is another constant, the value of $y(0)$.

Carrying out the integration yields

(7.14) $$y = \left(d - \frac{c}{r_2 - r_1}\right) e^{r_1 x} + \frac{c}{r_2 - r_1} e^{r_2 x}.$$

Regarding $[d - c/(r_2 - r_1]$ and $c/(r_2 - r_1)$ as new constants, c_1 and c_2, we see that any solution of Eq. (7.1) has the form

$$y = c_1 e^{r_1 x} + c_2 e^{r_2 x}, \tag{7.15}$$

where c_1 and c_2 are constants, provided, as mentioned above, that $r_1 \neq r_2$. Conversely, any function of the type appearing above satisfies Eq. (7.1). Since c and d can assume any values, we see that c_1 and c_2 can equally assume arbitrary values.

If $r_1 = r_2$, we see, upon returning to Eq. (7.13), that

$$y = de^{r_1 x} + cxe^{r_1 x}. \tag{7.16}$$

We have thus established the following important result.

Theorem. *Let r_1 and r_2 be the roots of the quadratic equation*

$$r^2 + ar + b = 0. \tag{7.17}$$

If $r_1 \neq r_2$, every solution of

$$y'' + ay' + by = 0 \tag{7.18}$$

has the form

$$y = c_1 e^{r_1 x} + c_2 e^{r_2 x}, \tag{7.19}$$

where c_1 and c_2 are constants.

If $r_1 = r_2$, every solution has the form

$$y = c_1 e^{r_1 x} + c_2 x e^{r_1 x}, \tag{7.20}$$

where again c_1 and c_2 are constants. Conversely, every function of the type appearing in (7.19) *or* (7.20) *satisfies an equation of the form* (7.18).

This solution is called the *general* solution.

In many cases, and indeed in most important cases, r_1 and r_2 are complex, which means that c_1 and c_2 are complex, if y is to be a real function. We shall discuss this point in Section 2.10.

2.8 AN ILLUSTRATIVE EXAMPLE

Consider the equation

$$y'' + 3y' + 2y = 0. \tag{8.1}$$

The characteristic equation is

$$r^2 + 3r + 2 = 0, \tag{8.2}$$

with the roots $r_1 = -2$, $r_2 = -1$. Hence the general solution is

$$y = c_1e^{-2x} + c_2e^{-x}. \tag{8.3}$$

Consider the equation

$$u'' + 2u' + u = 0. \tag{8.4}$$

The characteristic equation is

$$r^2 + 2r + 1 = 0 \tag{8.5}$$

with the roots -1, -1. Hence the general solution is

$$u = c_1e^{-x} + c_2xe^{-x}. \tag{8.6}$$

As we shall discuss in the next section, the constants c_1 and c_2 are determined by various conditions—initial conditions, boundary conditions, and so on—furnished by the underlying physical process.

2.9 THE EULER-DE MOIVRE FORMULAS

Before continuing our discussion, let us recall some important results connecting the trigonometric functions and the exponential function with a complex argument. The power-series expansions for e^{it}, $\cos t$, and $\sin t$ show that the following identity holds,

$$e^{it} = \cos t + i \sin t. \tag{9.1}$$

Similarly,

$$e^{-it} = \cos t - i \sin t. \tag{9.2}$$

These are the Euler-De Moivre formulas.

Adding and subtracting these relations, we have

$$\cos t = \frac{e^{it} + e^{-it}}{2}, \qquad \sin t = \frac{e^{it} - e^{-it}}{2i}. \tag{9.3}$$

EXERCISES

1. Use (9.3) to establish the relations $(\cos t)^2 = \frac{1}{2} + (\cos 2t)/2$, $(\sin t)^2 = \frac{1}{2} - (\cos 2t)/2$.
2. Similarly, use (9.3) to show that $\sin 2t = 2 \sin t \cos t$, $\cos 2t = \cos^2 t - \sin^2 t$.
3. Use the results of Exercise 1 to obtain the solution of $u' + u = (\cos t)^2$, $u(0) = 0$; of $u' - u = (\sin t)^2$, $u(0) = 1$.

2.10 REAL AND COMPLEX SOLUTIONS

In Section 2.7, we expressed the general solution of the differential equation (7.1) in terms of exponentials. These exponentials need not, however, correspond to real, or physical, solutions. Consider, for example, the very important equation

$$u'' + u = 0. \tag{10.1}$$

The characteristic equation is

$$r^2 + 1 = 0, \tag{10.2}$$

with the two distinct roots $r_1 = i$, $r_2 = -i$. The general solution is thus

$$u = c_1 e^{it} + c_2 e^{-it}, \tag{10.3}$$

where, as already indicated, c_1 and c_2 may be complex.

On the other hand, we know that $u = \cos t$, $u = \sin t$ are also solutions of Eq. (10.1). How do we explain this apparent overabundance of solutions? The answer lies in the formulas of the previous section, (9.2) and (9.3). Using these identities, we can convert complex solutions to real ones and back, depending on our needs and desires.

In dealing with derivatives and integrals, it is usually easier to employ the complex exponentials during calculations in order to take advantage of the relation

$$\frac{d}{dt}(e^{rt}) = re^{rt}, \tag{10.4}$$

valid for any r. After all of the manipulations have been performed, we can extract the desired real solution by taking real and complex parts. This important point will be discussed in Section 2.12.

Meanwhile, let us give a familiar example of this useful idea. Suppose that we wish to evaluate the integral

$$\int_0^x e^{ax_1} \cos bx_1 \, dx_1. \tag{10.5}$$

We begin with the simpler integral

$$\int_0^x e^{ax_1} e^{ibx_1} \, dx_1 = \frac{e^{(a+ib)x} - 1}{a + ib}. \tag{10.6}$$

Multiplying the numerator and denominator by $a - ib$, we get

$$\frac{(e^{(a+ib)x} - 1)(a - ib)}{(a + ib)(a - ib)} = \frac{a(e^{ax} \cos bx - 1) + be^{ax} \sin bx}{a^2 + b^2} + \frac{i[ae^{ax} \sin bx - b(e^{ax} \cos bx - 1)]}{a^2 + b^2}. \tag{10.7}$$

Equating real and complex parts, we obtain

$$\int_0^x e^{ax_1} \cos bx_1 \, dx_1 = \frac{a(e^{ax} \cos bx - 1) + be^{ax} \sin bx}{(a^2 + b^2)}, \tag{10.8}$$
$$\int_0^x e^{ax_1} \sin bx_1 \, dx_1 = \frac{ae^{ax} \sin bx - b(e^{ax} \cos bx - 1)}{a^2 + b^2}.$$

We shall use quite similar techniques in dealing with the equation of the electric circuit in Section 2.17.

EXERCISE

1. Show that if $a < 0$, then

$$\int_0^\infty e^{ax_1} \cos bx_1 \, dx_1 = \frac{-a}{a^2 + b^2}, \qquad \int_0^\infty e^{ax_1} \sin bx_1 \, dx_1 = \frac{b}{a^2 + b^2}.$$

2.11 THE EULER FORMALISM

Having observed the simple form of the general solution of

$$y'' + ay' + by = 0 \tag{11.1}$$

given in Section 2.7 it is natural to ask whether or not there is an easier and more direct way of obtaining it. It is to be expected that, once we have arrived at our objective, hindsight may furnish a simpler path.

Let us ask the following question. If we want e^{rx} to be a solution of Eq. (11.1), what value should the constant r have? Since

$$(e^{rx})' = re^{rx},$$
$$(e^{rx})'' = r^2e^{rx},$$

we see that

$$(e^{rx})'' + a(e^{rx})' + b(e^{rx}) = (r^2 + ar + b)e^{rx}. \tag{11.2}$$

Hence, if

$$r^2 + ar + b = 0, \tag{11.3}$$

then e^{rx} is a solution of Eq. (11.1). We recognize (11.3); it is the characteristic equation.

Hence, if r_1 and r_2 are the two roots of (11.3), assumed distinct for the moment, we have two distinct solutions of our differential equation.

EXERCISES

1. Find the general solution of each equation. If the characteristic roots are complex, write both the complex and real forms of the general solution.
   a) $y'' + y' = 0$ b) $3y'' + 14y' + 8y = 0$
   c) $y'' + 4y' + 4y = 0$ d) $y'' + y' + y = 0$
   e) $y'' - a^2y = 0$
2. In each part, write down an equation of the form $y'' + ay' + by = 0$ which has the following pair of solutions, or else tell why there can be no such equation.
   a) $1, x$ b) e^x, e^{3x}
   c) $\sin 2x, \cos 2x$ d) $e^{2x} \sin x, e^{2x} \cos x$
   e) e^{-x}, xe^{-x} f) x, x^2

2.12 LINEARITY AND THE PRINCIPLE OF SUPERPOSITION

Why is it useful to find particular solutions of differential equations? It is important in this case because the equation we are investigating is *linear* in the function and in its derivatives. From this we can conclude that the sum of two solutions is again a solution. Let us see how this comes about. Let y_1 and y_2 be two solutions of

$$y'' + ay' + by = 0, \tag{12.1}$$

which is to say

$$\begin{aligned} y_1'' + ay_1' + by_1 &= 0, \\ y_2'' + ay_2' + by_2 &= 0. \end{aligned} \tag{12.2}$$

Adding the two equations, we have

$$y_1'' + y_2'' + a(y_1' + y_2') + b(y_1 + y_2) = 0 \tag{12.3a}$$

or

$$(y_1 + y_2)'' + a(y_1 + y_2)' + b(y_1 + y_2) = 0, \tag{12.3b}$$

which means that $y_1 + y_2$ is also a solution.

Furthermore, we note that if y_1 is a solution, then c_1y_1 is a solution for any constant c_1. This follows as above. From (12.2) we have

$$c_1y_1'' + c_1ay_1' + c_1by_1 = 0 \tag{12.4a}$$

or

$$(c_1y_1)'' + a(c_1y_1)' + b(c_1y_1) = 0, \tag{12.4b}$$

which means that c_1y_1 is also a solution.

We have thus established the fundamental result that the linear combination $c_1y_1 + c_2y_2$ is a solution of Eq. (12.1) for any constants c_1 and c_2 whenever y_1 and y_2 are solutions. This is the basic technique of *superposition*, the cornerstone of analysis. Much of advanced mathematics, and certainly of partial differential equations and mathematical physics, is based on this simple, yet amazingly powerful, idea.

Why didn't we employ this procedure from the beginning to obtain the general solution $y = c_1e^{r_1x} + c_2e^{r_2x}$ instead of the circuitous route we followed? There are two parts to the answer. In the first place, as we shall indicate below, we can obtain the general solution of the important equation

$$y'' + ay' + by = g(x) \tag{12.5}$$

using the first method given in the preceding pages. In the second place, we would still have had to wonder and worry about the possibility of other types of solutions, i.e., about uniqueness of solution.

Once we have established uniqueness, we are free to follow any convenient route to the solution.

EXERCISES

1. By considering $u' = u^2 - 1$ show that, in general, superposition does not hold for nonlinear equations. Here $u_1 = 1$, $u_2 = -1$ are two particular solutions. But $u_1 + u_2 = 0$ is not a solution.

2. If $r_2 = r_1$, we can deduce the existence of a second solution of the form xe^{r_1x} by means of the following argument. If $r_2 \neq r_1$, then the foregoing discussion assures us that

$$y = \frac{e^{r_2x} - e^{r_1x}}{r_2 - r_1}$$

is a solution. The limiting expression as $r_2 \to r_1$ is precisely xe^{r_1x}.

3. Exercise 2 is suggestive but not rigorous without some further considerations. Show directly that if

$$r^2 + ar + b = 0$$

has r_1 as a repeated root, then xe^{r_1x} is a solution of Eq. (12.1).

4. Establish Euler's relations using the fact that the particular solutions $\cos x$ and $\sin x$ of Eq. (10.1) must be linear combinations of e^{ix} and e^{-ix} and then matching the values at $x = 0$.

5. Consider the equation $y'' + y = 0$. The general solution may be written as

$$y = c_1 \cos x + c_2 \sin x.$$

Since $\sin (x + b)$ is a solution for any constant b, we must have

$$\sin (x + b) = c_1 \cos x + c_2 \sin x.$$

Equating the values of the function and its derivative at $x = 0$, we obtain the addition formula

$$\sin (x + b) = \sin b \cos x + \cos b \sin x.$$

Similarly, show that

$$\cos (x + b) = \cos x \cos b - \sin x \sin b.$$

6. Given $y'' + y = 0$, show directly that

$$\frac{d}{dx}(y'^2 + y^2) = 0$$

and thus $y'^2 + y^2 = c$, a constant, for any solution y. Hence show that

$$\sin^2 x + \cos^2 x = 1.$$

(The point we are trying to establish by means of Exercises 4, 5, and 6 is that many of the fundamental properties of the basic functions of analysis can be demonstrated directly with a minimum of calculation using the defining differential equation.)

7. Given $y'' + y - \frac{1}{6}y^3 = 0$, show that

$$\frac{d}{dx}(y'^2 + y^2 - \tfrac{1}{12}y^4) = 0$$

and thus $y'^2 + y^2 - y^4/12 = c$, a constant, for any solution y.

8. Given $y'' + \sin y = 0$, show that

$$\frac{d}{dx}(y'^2 - 2\cos y) = 0$$

and thus $y'^2 - 2\cos y = c$, a constant, for any solution y.

9. In all three cases above (Exercises 6, 7, 8) determine the values of y for which $|y|$ is a maximum. Determine the slopes at the points where $y = 0$.

2.13 THE RICCATI EQUATION

In general, differential equations containing nonlinear terms are extraordinarily difficult to analyze. This is why we concentrate on linear equations in an introductory course. There is, however, one class of nonlinear equations of great importance and frequent occurrence, where the general solution can readily be obtained. This is the Riccati equation

$$z' + z^2 + az + b = 0. \tag{13.1}$$

The reason for this is that the solution of (13.1) may be written in the form

$$z = y'/y, \tag{13.2}$$

where y satisfies the second-order linear differential equation

$$y'' + ay' + by = 0. \tag{13.3}$$

We can easily verify this directly. From (13.2) we have

$$y' = yz, \qquad (13.4)$$
$$y'' = (yz)' = y'z + yz'$$
$$= yz^2 + yz' = y(z^2 + z').$$

Hence, if (13.3) holds we have

$$y(z' + z^2) + ayz + by = 0, \qquad (13.5)$$
$$y[z' + z^2 + az + b] = 0.$$

Thus, where $y \neq 0$, z satisfies (13.1).

Examining the proof we see that we have not used the fact that a and b are constants. Hence, the result holds for the case where a and b are functions of the independent variable.

EXERCISES

1. Find the general solution of
   a) $z' + z^2 + 1 = 0$,
   b) $z' + z^2 - 1 = 0$,
   c) $z' + z^2 + z - 1 = 0$.

2. What is the general solution of $z' + cz^2 + az + b = 0$, where a, b, c are constants?

2.14 INITIAL-VALUE PROBLEMS

As we hope to have indicated in Chapter 1 by means of the illustrative examples, one of the most important uses of the theory of differential equations is that of determining the future behavior of a physical system, given the present state.

In mathematical terms, this means that we have a differential equation such as

$$y'' + ay' + by = 0, \qquad (14.1)$$

with precisely enough information concerning y at $x = 0$ to determine the arbitrary constants c_1 and c_2 appearing in the general solution, no more and no less. Let us see how this goes.

Suppose that the initial data were

$$y(0) = a_1, \qquad y'(0) = a_2. \tag{14.2}$$

Since the general solution has the form

$$y = c_1 e^{r_1 x} + c_2 e^{r_2 x} \tag{14.3}$$

(considering first the case of distinct characteristic roots), the conditions (14.2) lead to the following two linear algebraic equations in c_1 and c_2:

$$a_1 = c_1 + c_2, \qquad a_2 = r_1 c_1 + r_2 c_2. \tag{14.4}$$

Since the determinant for these two equations is

$$\begin{vmatrix} 1 & 1 \\ r_1 & r_2 \end{vmatrix} = r_2 - r_1, \tag{14.5}$$

which is not equal to zero by assumption, we know that (14.4) has a unique solution:

$$c_1 = \frac{a_1 r_2 - a_2}{r_2 - r_1}, \qquad c_2 = \frac{a_2 - a_1 r_1}{r_2 - r_1}. \tag{14.6}$$

If r_1 is a repeated root, the algebra is even simpler. The general solution then has the form

$$y = c_1 e^{r_1 x} + c_2 x e^{r_1 x}. \tag{14.7}$$

Use of the initial conditions (14.2) leads to the equations

$$a_1 = c_1, \qquad a_2 = c_1 r_1 + c_2, \tag{14.8}$$

which obviously possess a unique solution.

Theorem. *The equation*

$$y'' + ay' + by = 0, \tag{14.9}$$

subject to the initial conditions

$$y(0) = a_1, \qquad y'(0) = a_2, \tag{14.10}$$

possesses a unique solution of either the form (14.3) *or* (14.7), *where* c_1 *and* c_2 *are determined as in* (14.6) *or* (14.8), *respectively.*

EXERCISES

1. Find the solution of each equation which satisfies the given initia conditions:

   a) $y'' + y' = 0,\quad y(-1) = 1,\quad y'(-1) = 3,$
   b) $3y'' + 14y' + 8y = 0,\quad y(0) = 4,\quad y'(0) = -6,$
   c) $y'' + 4y' + 4y = 0,\quad y(0) = -2,\quad y'(0) = \frac{9}{2},$
   d) $y'' + y' + y = 0,\quad y(0) = 2,\quad y'(0) = 2,$
   e) $y'' - a^2y = 0,\quad y(1) = e^a,\quad y'(1) = ae^a.$

2. Let y_1 be the solution of
$$y'' + ay' + by = 0, \qquad y(0) = 1, \qquad y'(0) = 0,$$
and y_2 be the solution of
$$y'' + ay' + by = 0, \qquad y(0) = 0, \qquad y'(0) = 1.$$
These are often called the *principal solutions.* Show that the solution of
$$y'' + ay' + by = 0, \qquad y(0) = a_1, \qquad y'(0) = a_2,$$
may be written as $y = a_1y_1 + a_2y_2$.

3. Determine the explicit forms of y_1 and y_2 in Exercise 2 first in the case where $r_1 \neq r_2$, and then where $r_1 = r_2$.

4. Show that if $a > 0$, $b > 0$, then all solutions of
$$y'' + ay' + by = 0$$
approach zero as $x \to \infty$.

5. Consider the equation
$$y'' + y = 0, \qquad y(0) = 1, \qquad y'(0) = 0.$$
We have $y = \cos x$, $y' = -\sin x$. Consider y and y' as the coordinates of point P in the yy'-plane, called the *phase plane* (see Fig. 2.5). Show that as x increases from 0 to 2π, the point P traverses a circle $y^2 + y'^2 = 1$.

6. Show that if
$$y'' + y = 0, \qquad y(0) = c_1, \qquad y'(0) = c_2,$$
then $P = (y, y')$ traverses a circle of radius $(c_1^2 + c_2^2)^{1/2}$ with center at the origin.

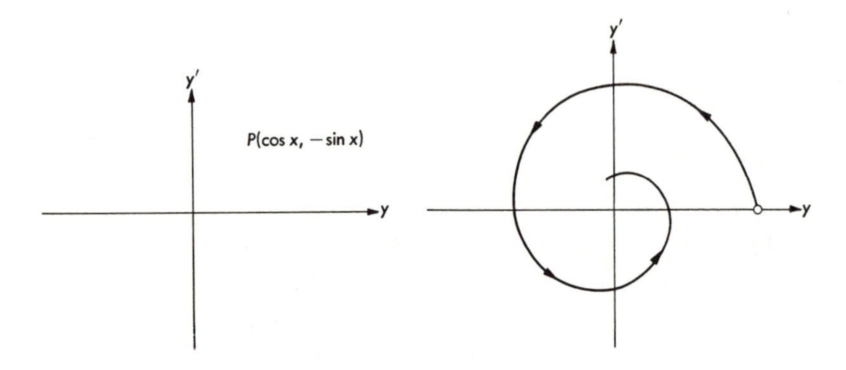

Figure 2.5 **Figure 2.6**

7. Show that if

$$y'' + b^2y = 0, \qquad y(0) = 1, \qquad y'(0) = 0,$$

then P traverses an ellipse.

8. If $y'' + 3y' + 2y = 0$, what possible types of curves can $P = (y, y')$ describe?

9. If $y'' + y' + y = 0$, $y(0) = 1$, $y'(0) = 0$, show that P traces out a spiral, as shown in Fig. 2.6.

10. Let C_1 denote the curve traced by (y, y'), the solution of

$$y'' + ay' + by = 0, \qquad y(0) = a_1, \qquad y'(0) = b_1,$$

and C_2 denote the corresponding curve when the initial conditions are $y(0) = a_2$, $y'(0) = b_2$. Can C_1 and C_2 intersect in the phase plane?

2.15 AN *RC*-CIRCUIT

Let us now illustrate the foregoing results by means of some types of simple but extremely important electrical circuits. What is amazing is the number of interesting phenomena which can be quite accurately described by an equation of such simple form as Eq. (14.9).

We begin with a circuit containing a voltage source, a resistor, and a capacitor. The switch is closed at time $t = 0$. How does the current behave as a function of time? (See Fig. 2.7.) The resistance R is measured in ohms and the capacitance C in farads. Their order of magnitude is of no importance at the moment. To avoid confusion with $i = \sqrt{-1}$, let us denote the current

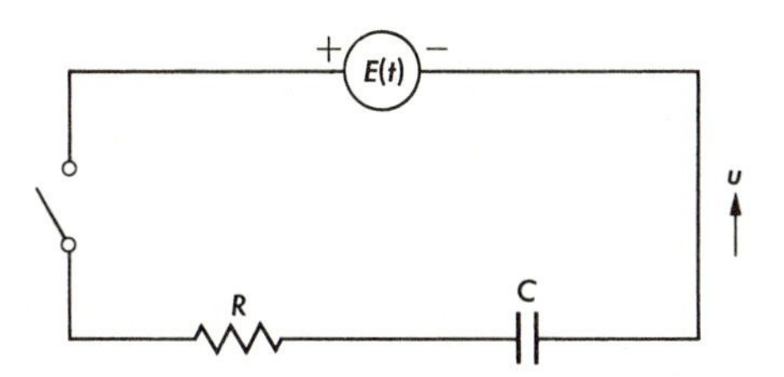

Figure 2.7

by $u(t)$. Then the balancing of voltages in the circuit yields the equation

$$Ru + \frac{1}{C}\int_0^t u\,dt_1 = E(t). \tag{15.1}$$

Let us consider first the case where $E(t)$ is constant so that Eq. (15.1) has the form

$$Ru + \frac{1}{C}\int_0^t u\,dt_1 = E_0. \tag{15.2}$$

There are two approaches we can use. One is to differentiate and thus reduce Eq. (15.2) to a differential equation

$$R\frac{du}{dt} + \frac{u}{C} = 0. \tag{15.3}$$

The other is to regard $\int_0^t u\,dt_1$ as a new variable,

$$v = \int_0^t u\,dt_1, \tag{15.4}$$

and write (15.2) as

$$R\frac{dv}{dt} + \frac{v}{C} = E_0, \qquad v(0) = 0. \tag{15.5}$$

Let us consider Eq. (15.5) first. By using the result of Section 2.3, we have

$$v = e^{-t/RC}\int_0^t e^{t_1/RC}\frac{E_0}{R}\,dt_1 = E_0C(1 - e^{-t/RC}). \tag{15.6}$$

Hence

$$u = \frac{dv}{dt} = \frac{E_0}{R}e^{-t/RC}. \tag{15.7}$$

Using (15.3) with the initial condition $Ru(0) = E_0$, obtained from Eq. (15.2), we arrive at precisely the same result. We see that the current decreases to

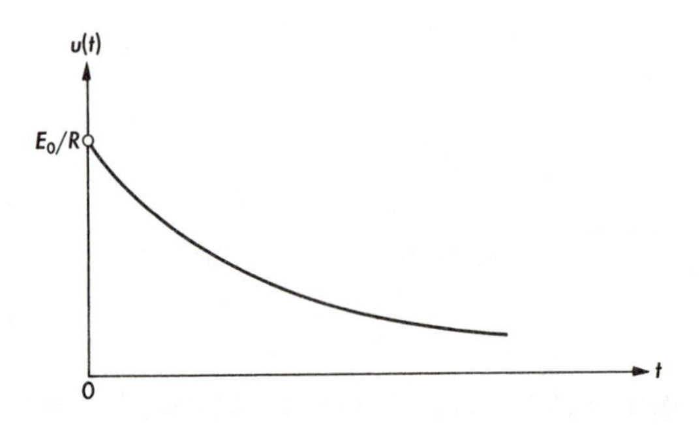

Figure 2.8

zero as $t \to \infty$. A capacitor represents an open circuit so far as direct current is concerned. (See Fig. 2.8.)

EXERCISES

1. How long does it take the current to decrease to one-half of its initial value? This is an important constant associated with the RC-circuit, the "time constant."
2. Given E_0 and the values of $u(t)$ at two points t_1, t_2, $0 < t_1 < t_2$, can one always determine the values of the circuit constants?

2.16 $E(t) = E_0 \cos \omega t$

If the voltage is alternating so that

$$E(t) = E_0 \cos \omega t,$$

then in place of Eq. (15.2) we obtain

$$Ru + \frac{1}{C}\int_0^t u\, dt_1 = E_0 \cos \omega t, \tag{16.1}$$

which may be written [cf. Eq. (15.4)], as

$$R\frac{dv}{dt} + \frac{v}{C} = E_0 \cos \omega t, \qquad v(0) = 0. \tag{16.2}$$

Referring to Section 2.3, we see that the solution to this equation has the form

$$\begin{aligned} v &= e^{-t/RC}\int_0^t \frac{E_0}{R} e^{t_1/RC} \cos \omega t_1\, dt_1 \\ &= \frac{E_0}{R}\left[\frac{1/RC(\cos \omega t - e^{-t/RC}) + \omega \sin \omega t}{1/(RC)^2 + \omega^2}\right]. \end{aligned} \tag{16.3}$$

Observe that the solution consists of two parts. The first part,

$$v_1 = \frac{-E_0/R(e^{-t/RC}/RC)}{1/(RC)^2 + \omega^2}, \tag{16.4}$$

decreases in amplitude as t increases, and eventually becomes arbitrarily small. Its effect on the current ultimately disappears. The second part,

$$v_2 = \frac{E_0}{R}\left[\frac{1/RC \cos \omega t + \omega \sin \omega t}{1/(RC)^2 + \omega^2}\right], \tag{16.5}$$

is periodic in t, with the same period and thus frequency as the applied voltage.

The first term, (16.4), representing an effect which ultimately disappears as time increases, is called a *transient* term. The second term, which persists with time, and indeed is periodic, is called the *steady-state* term. In some processes, the steady-state effect is of major significance, in others the transient. Whether the transient or the steady-state term is the more important depends on the time scale, which is to say the length of time the system is observed subsequent to the initial instant.

It is important to note that the transient effect can be quite disastrous. Consider the current $u = dv/dt$, and suppose that R is quite small. Then the term (16.5) yields

$$\begin{aligned} u_2 = \frac{dv_2}{dt} &= \frac{E_0}{R}\left[\frac{-\omega/RC \sin \omega t + \omega^2 \cos \omega t}{1/(RC)^2 + \omega^2}\right] \\ &= \frac{E_0}{R}\left[\frac{-\omega RC \sin \omega t + \omega^2 (RC)^2 \cos \omega t}{1 + (RC)^2\omega^2}\right] \\ &\cong \frac{E_0}{R}\left[\frac{-\omega RC \sin \omega t}{1}\right] = -\omega C E_0 \sin \omega t, \end{aligned} \tag{16.6}$$

where the symbol $\cong$ is to be read "is approximately equal to." The transient term (16.4), however, yields

(16.7)

$$u_1 = \frac{dv_1}{dt} = \frac{E_0/R(e^{-t/RC})/(RC)^2}{1/(RC)^2 + \omega^2} = \frac{E_0}{R}\frac{e^{-t/RC}}{1 + (RC)^2\omega^2} \cong \frac{E_0}{R}e^{-t/RC}.$$

If t is small, for example $t = RC$, we see that

$$u_1 \cong \frac{E_0}{R}e^{-1}, \tag{16.8}$$

a quantity which can be quite large if R is small. This explains one way in which fuses can blow when we flip a switch.

2.17 $E(t) = E_0 \cos \omega t$ (SUPERPOSITION)

In order to solve Eq. (16.2), we had to evaluate a complicated integral given by (16.3). Can we simplify the calculations a bit? The equation.

$$R\frac{dv}{dt} + \frac{v}{C} = E_0 \cos \omega t, \qquad v(0) = 0, \tag{17.1}$$

may be written

$$R\frac{dv}{dt} + \frac{v}{C} = E_0\left(\frac{e^{i\omega t} + e^{-i\omega t}}{2}\right), \qquad v(0) = 0. \tag{17.2}$$

We maintain that $v = w_1 + w_2$, where w_1 is the solution of

$$R\frac{dw_1}{dt} + \frac{w_1}{C} = \frac{E_0 e^{i\omega t}}{2}, \qquad w_1(0) = 0, \tag{17.3}$$

and w_2 is the solution of

$$R\frac{dw_2}{dt} + \frac{w_2}{C} = \frac{E_0 e^{-i\omega t}}{2}, \qquad w_2(0) = 0. \tag{17.4}$$

The proof is immediate. Adding Eqs. (17.3) and (17.4), we obtain

$$R\left(\frac{dw_1}{dt} + \frac{dw_2}{dt}\right) + \frac{1}{C}(w_1 + w_2) = E_0\left(\frac{e^{i\omega t} + e^{-i\omega t}}{2}\right), \tag{17.5}$$

$$R\frac{d}{dt}(w_1 + w_2) + \frac{1}{C}(w_1 + w_2) = E_0\left(\frac{e^{i\omega t} + e^{-i\omega t}}{2}\right).$$

This is taking every advantage of the linearity of the equation. In general, then, if we put two voltage sources, $E_1(t)$ and $E_2(t)$, in the circuit, as indicated in Fig. 2.9, then the current u flowing in the circuit is equal to $u_1 + u_2$, where u_1 is the current that would flow if E_1 alone were present and u_2 is the current that would flow if E_2 alone were present.

This is another remarkable superposition property of linear differential equations.

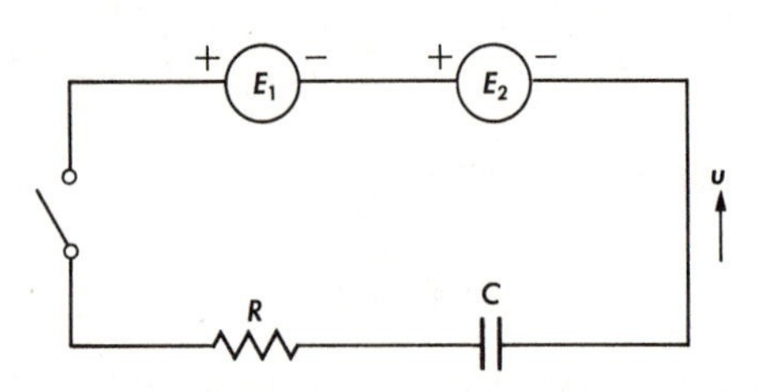

Figure 2.9

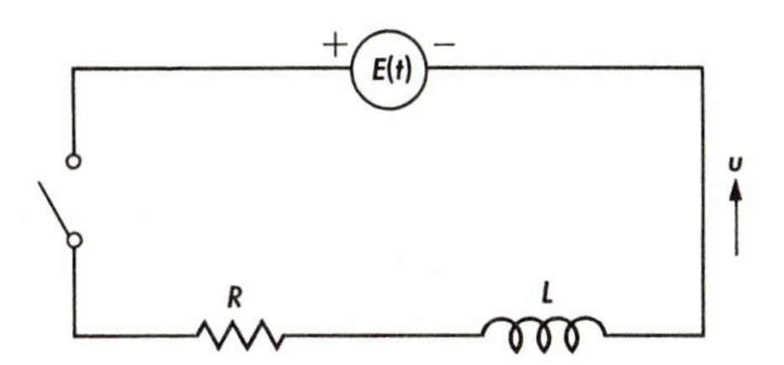

Figure 2.10

To solve Eqs. (17.3) and (17.4), we proceed as above, with the aid of simpler integrals such as

$$\int_0^t e^{-t_1/RC} e^{i\omega t_1}\, dt_1, \tag{17.6}$$

2.18 AN *RL*-CIRCUIT

Consider now the situation where an inductor and a resistor are present in a circuit (Fig. 2.10). The equation governing the system is then

$$L\frac{du}{dt} + Ru = E_0, \qquad u(0) = 0. \tag{18.1}$$

The initial condition is obtained from the fact that the current in the system is zero at the instant when the switch is closed. The solution of Eq. (18.1) is

$$\begin{aligned} u &= e^{-Rt/L} \int_0^t \frac{E_0}{L} e^{Rt_1/L}\, dt_1 \\ &= \frac{E_0}{R} - \frac{E_0}{R} e^{-Rt/L}. \end{aligned} \tag{18.2}$$

Observe that here the solution approaches the steady-state value E_0/R as $t \to \infty$. (See Fig. 2.11.)

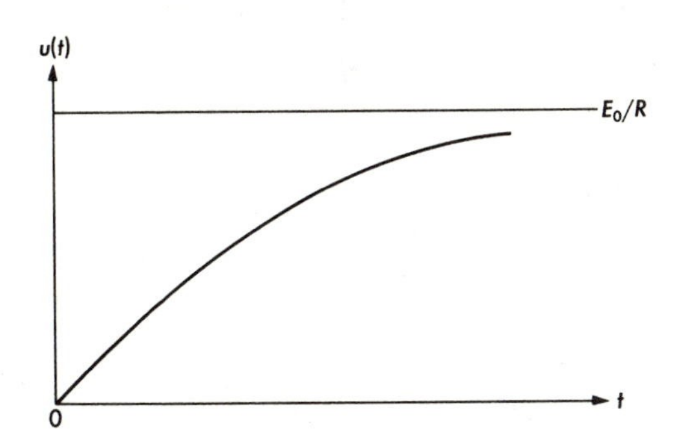

Figure 2.11

EXERCISES

1. Obtain the solution for

$$L\frac{du}{dt} + Ru = E_0 \cos \omega t, \qquad u(0) = 0.$$

What is the steady-state solution?

2. If we observe the value of $u(t)$ at two instants, $t_1, t_2, t_2 > t_1 > 0$, can we determine the values of R and L?
3. Given the observations at t_1 and t_2, and the value of R or L, can we determine the values of the missing circuit parameter and the forcing frequency ω?

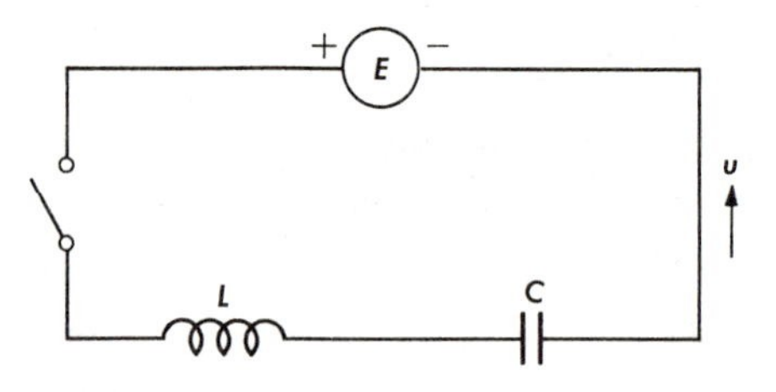

Figure 2.12

2.19 AN LC-CIRCUIT

The next circuit, one containing an inductor and capacitor, but no resistor, possesses a very interesting property, as we shall see in a moment. (See Fig. 2.12.) Our basic equation is now

$$L\frac{du}{dt} + \frac{1}{C}\int_0^t u\,dt_1 = E(t). \tag{19.1}$$

Consider first the case where $E(t) = E_0$, a constant:

$$L\frac{du}{dt} + \frac{1}{C}\int_0^t u\,dt_1 = E_0. \tag{19.2}$$

Differentiating, we have

$$L\frac{d^2u}{dt^2} + \frac{u}{C} = 0, \tag{19.3}$$

with the initial conditions

$$u(0) = 0, \qquad Lu'(0) = E_0. \tag{19.4}$$

The first is the familiar statement that the initial current is zero, and the second follows from Eq. (19.2) upon setting $t = 0$.

The general solution of Eq. (19.3) is

$$u = c_1 \cos (t/\sqrt{LC}) + c_2 \sin (t/\sqrt{LC}). \tag{19.5}$$

Using the first condition in (19.4), we see that $c_1 = 0$; the second yields

$$Lc_2/\sqrt{LC} = E_0, \qquad c_2 = E_0\sqrt{C/L}. \tag{19.6}$$

Hence the expression for the current is

$$u = E_0\sqrt{C/L} \sin (t/\sqrt{LC}). \tag{19.7}$$

This represents a periodic phenomenon with period equal to $2\pi\sqrt{LC}$. The behavior of this circuit is quite different from that of either the *RC*- or *RL*-circuits. (See Fig. 2.13.)

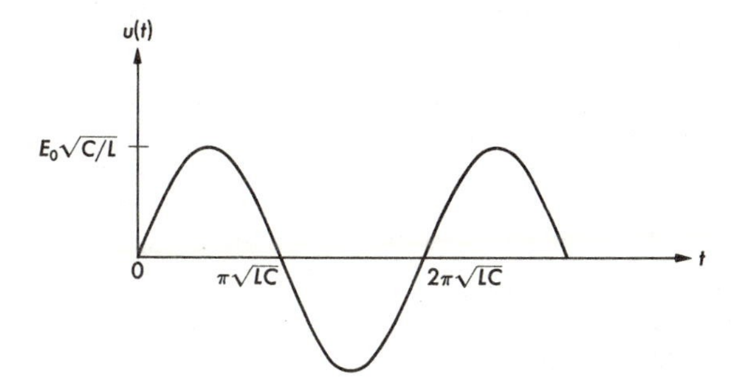

Figure 2.13

2.20 AN *LC*-CIRCUIT WITH ALTERNATING VOLTAGE

What happens when $E(t) = E_0 \cos \omega t$? The equation for the current is now

$$L\frac{du}{dt} + \frac{1}{C}\int_0^t u\, dt_1 = E_0 \cos \omega t. \tag{20.1}$$

Differentiation yields

$$L\frac{d^2u}{dt^2} + \frac{u}{C} = -E_0\omega \sin \omega t, \tag{20.2}$$

an equation whose solution escapes any of our previous techniques. Setting

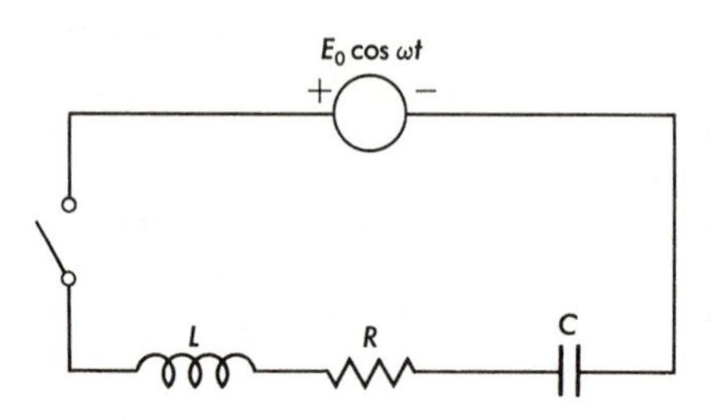

Figure 2.14

$v = \int_0^t u \, dt_1$ is of no help since we obtain

(20.3)
$$L \frac{d^2 v}{dt^2} + \frac{v}{C} = E_0 \cos \omega t,$$

an equation of the same type as above.

In order to obtain the complete solution of Eq. (20.2), we must introduce some new ideas. As long as we are going to all of this effort, we may just as well consider the general *LRC*-circuit with periodic voltage (see Fig. 2.14). The equation for the current is now

(20.4)
$$L \frac{du}{dt} + Ru + \frac{1}{C} \int_0^t u \, dt_1 = E_0 \cos \omega t,$$

yielding upon differentiation the differential equation

(20.5)
$$L \frac{d^2 u}{dt^2} + R \frac{du}{dt} + \frac{u}{C} = -E_0 \omega \sin \omega t,$$

subject to the initial conditions

(20.6)
$$u(0) = 0, \qquad Lu'(0) = E_0.$$

A consideration of this equation enables us to introduce some ideas of fundamental importance in the theory of linear differential equations. The principal concept is, as before, the principle of superposition.

2.21 LINEAR HOMOGENEOUS AND INHOMOGENEOUS EQUATIONS

So far we have studied and resolved the problem of obtaining the complete solution of the differential equation

(21.1)
$$y'' + ay' + by = 0.$$

This is a linear equation because of the fact that all the derivatives occur separately and to the first power; it is called *homogeneous* because y and ky, for any constant k, are simultaneously solutions for the equation. We have already made essential use of this fact in solving the initial-value problem.

An equation of the foregoing type can be considered to describe the behavior of a system subject to no external influences. In many important cases, however, the system is exposed to external forces, as in the case discussed in Section 2.20 where the alternating voltage plays this role. The differential equation describing the system now has the form

$$y'' + ay' + by = f(x). \tag{21.2}$$

This is called an *inhomogeneous* equation. If $f(x)$ is not equal to zero, y and ky are not simultaneously solutions. Obviously, it is desirable to reduce the solution of an inhomogeneous equation to that of a homogeneous equation, since equations of the latter type possess attractive features.

To accomplish this reduction, we use the linearity of the left-hand side of Eq. (21.2). Let z denote some solution of (21.2), a *particular* solution. Then

$$z'' + az' + bz = f(x). \tag{21.3}$$

Let us subtract Eq. (21.3) from (21.2). We get

$$y'' - z'' + a(y' - z') + b(y - z) = 0, \tag{21.4}$$

which may be written

$$(y - z)'' + a(y - z)' + b(y - z) = 0. \tag{21.5}$$

What does this equation say? It states that $y - z$ is a solution of the homogeneous equation. Hence we can assert the following:

Theorem. *The general solution of Eq.* (21.2) *may be obtained by adding a particular solution of* (21.2) *to the general solution of Eq.* (21.1).

Everything now hinges on obtaining a particular solution of (21.2). Fortunately, in a number of important cases, this is quite easy to do.

EXERCISES

---

1. If y_1 is a particular solution of

$$y'' + ay' + by = f_1(t),$$

and y_2 is a particular solution of

$$y'' + ay' + by = f_2(t),$$

then

$$y = y_1 + y_2$$

is a particular solution of

$$y'' + ay' + by = f_1(t) + f_2(t).$$

2. If a and b are real and y_1 is a particular solution of

$$y'' + ay' + by = e^{i\omega t},$$

then the real part of y_1 is a particular solution of

$$y'' + ay' + by = \cos \omega t,$$

and the complex part of y_1 is a particular solution of

$$y'' + ay' + by = \sin \omega t.$$

2.22 ILLUSTRATIVE EXAMPLES

Consider the equation

$$y'' + 4y = 1. \tag{22.1}$$

A particular solution is easily seen to be $y = \frac{1}{4}$. Hence the general solution is

$$y = \tfrac{1}{4} + c_1 \cos 2x + c_2 \sin 2x. \tag{22.2}$$

Consider the equation

$$u'' - u = e^{2t}. \tag{22.3}$$

A particular solution is seen to be $\frac{1}{3}e^{2t}$. Hence the general solution is

$$u = c_1 e^t + c_2 e^{-t} + \tfrac{1}{3}e^{2t}. \tag{22.4}$$

Suppose that we are required to solve the initial-value problem

$$y'' + 4y = 1, \qquad y(0) = 0, \qquad y'(0) = 1. \tag{22.5}$$

Using the general solution of (22.2), we obtain the equations

$$0 = \tfrac{1}{4} + c_1, \qquad 1 = 2c_2. \tag{22.6}$$

Hence the solution of Eq. (22.5) is given by

$$y = \tfrac{1}{4} - \tfrac{1}{4}\cos 2x + \tfrac{1}{2}\sin 2x. \tag{22.7}$$

EXERCISES

1. Find the solution of $u'' - u = e^{2t}$ satisfying the initial conditions $u(0) = 1$, $u'(0) = 0$.
2. Find the solution of

$$L\frac{d^2v}{dt^2} + \frac{v}{C} = E, \qquad v(0) = 0, \qquad v'(0) = 0.$$

2.23 EXPONENTIAL FORCING TERM

It is all very well to guess particular solutions, but it is far better to possess a systematic approach. Ingenuity is the last resort of a mathematician when he has no recourse to a powerful general theory. For the equation

$$y'' + ay' + by = e^{cx}, \tag{23.1}$$

we are fortunate in possessing a very simple direct method. We begin with the observation, already made, that the result of differentiating e^{cx} is to multiply by the constant c. Thus

$$(e^{cx})' = ce^{cx}, \tag{23.2}$$

$$(e^{cx})'' = c^2e^{cx}.$$

Hence the function $y = ke^{cx}$ will be a solution of Eq. (23.1) provided the constant k is chosen properly. Substituting in (23.1), we have

$$(c^2 + ac + b)ke^{cx} = e^{cx}. \tag{23.3}$$

Hence the appropriate choice of k is

$$k = \frac{1}{c^2 + ac + b}, \tag{23.4}$$

provided that the denominator is not zero. We will discuss this possibility subsequently.

We see now how the particular solution to Eq. (22.3) was obtained.

EXERCISES

1. Find a particular solution and the general solution of each of the following equations.
   a) $y'' + y' = \sin 2t$
   b) $3y'' + 14y' + 8y = e^{-x}$
   c) $y'' + 4y' + 4y = e^t$
   d) $y'' + y' + y = \cos t$
   e) $y'' - a^2y = \sin \omega t$

2.24 ALTERNATING VOLTAGE, CONTINUED

Armed with the technique given in Section 2.23, let us return to the study of the equation first encountered in Section 2.20:

$$L\frac{d^2u}{dt^2} + R\frac{du}{dt} + \frac{u}{C} = -E_0\omega \sin \omega t. \tag{24.1}$$

We first consider the equation

$$L\frac{d^2u}{dt^2} + R\frac{du}{dt} + \frac{u}{C} = e^{i\omega t}. \tag{24.2}$$

A particular solution, following the procedure outlined in the previous section, is

$$u = \frac{e^{i\omega t}}{(-L\omega^2 + i\omega R + 1/C)}. \tag{24.3}$$

To obtain a solution for Eq. (24.1), we use the fact that

$$\sin \omega t = \frac{e^{i\omega t} - e^{-i\omega t}}{2i}. \tag{24.4}$$

Hence, using (24.3) for ω and $-\omega$, we find as a particular solution of (24.1)

$$u = -\frac{E_0\omega}{2i}\left[\frac{e^{i\omega t}}{(-L\omega^2 + i\omega R + 1/C)} - \frac{e^{-i\omega t}}{(-L\omega^2 - i\omega R + 1/C)}\right]. \tag{24.5}$$

Rationalizing, we have

(24.6)

$$u = -\frac{E_0\omega}{2i}\left[\frac{e^{i\omega t}(-L\omega^2 - i\omega R + 1/C) - e^{-i\omega t}(-L\omega^2 + i\omega R + 1/C)}{(-L\omega^2 + 1/C)^2 + \omega^2R^2}\right]$$

$$= -\frac{E_0\omega}{2i}\left[\frac{(1/C - L\omega^2)(e^{i\omega t} - e^{-i\omega t}) - i\omega R(e^{i\omega t} + e^{-i\omega t})}{(-L\omega^2 + 1/C)^2 + \omega^2R^2}\right]$$

$$= \frac{-E_0\omega(1/C - L\omega^2)\sin\omega t + E_0R\omega^2\cos\omega t}{(-L\omega^2 + 1/C)^2 + \omega^2R^2}.$$

With the aid of this particular solution, representing a steady-state periodic current, we can obtain the general solution of Eq. (24.1) by adding a term of the form $c_1e^{r_1t} + c_2e^{r_2t}$, which is the general solution of

$$L\frac{d^2u}{dt^2} + R\frac{du}{dt} + \frac{u}{C} = 0, \tag{24.7}$$

assuming, as is the case in most important situations, that $r_1 \neq r_2$. Since $R, L, C > 0$ in electrical circuits, the general solution for Eq. (24.7) always represents the transient term. If R is small, this transient term may, however, persist for quite a long time.

An important point to note is that the long-term behavior of an *LRC*-circuit with an alternating voltage is independent of the initial conditions.

2.25 NO RESISTANCE

Let us consider the case where the resistance is taken to be zero. This is, of course, always an idealization, but it represents a useful way of considering the case where the resistance is quite small. (See Fig. 2.15.) The general solution for this system consists of two parts, a particular solution with

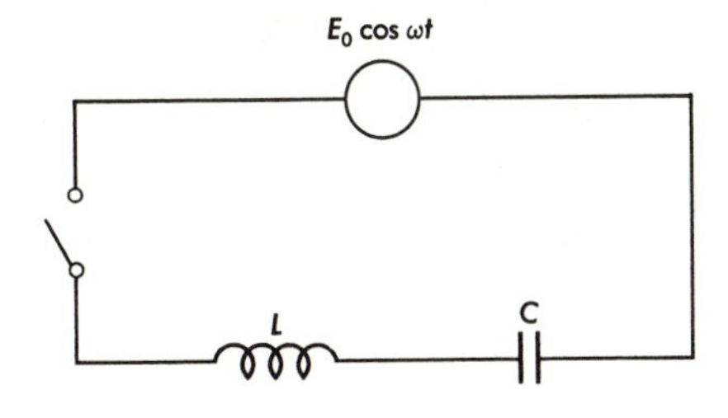

Figure 2.15

frequency ω and a general solution of Eq. (19.3) with frequency $1/\sqrt{LC}$. We suppose for the moment that these frequencies are different. If $R = 0$ and $\omega = 1/\sqrt{LC}$, the particular solution of the form (24.6) is meaningless. What happens in this case is discussed in Section 2.26.

The frequency $1/\sqrt{LC}$ is called the *natural frequency* of the circuit. What the foregoing states is that a periodic superimposed influence generates a corresponding influence of the same period in the system, the steady-state solution, together with another influence, with a period determined by the intrinsic properties of the system. In many cases, this second part of the solution is highly undesirable. It may result in loss of energy, overheating, or, in mechanical systems, dangerous vibrations. We shall discuss this point in the following section.

A number of interesting phenomena in nature can be explained in terms of the behavior of a sum of two periodic disturbances, but any discussion of this type would take us too far into the domain of physics.

2.26 RESONANCE

What happens if ω assumes the forbidden value $1/\sqrt{LC}$? In this case, formula (24.6) cannot be used. Let us then look for a particular solution of

$$Lu'' + \frac{u}{C} = e^{i\omega t} \tag{26.1}$$

under these new circumstances. Here $\omega = 1/\sqrt{LC}$. It will simplify all of the algebra which follows if we consider the special case $L = C = \omega = 1$, which illustrates the phenomenon of the general case. Consider then

$$u'' + u = e^{it}. \tag{26.2}$$

There are two ways we can proceed. We can derive the general solution of this equation using the factorization method of Section 2.7, or we can "guess" a solution. This "guess" is, of course, based on our knowledge of the actual solution. It can be justified in terms of the kind of argument given in Exercise 2 of Section 2.12.

Let us suppose that we are ingenious enough to think of looking for a particular solution of the form

$$u = ate^{it}, \tag{26.3}$$

where a is a constant to be determined. Then

(26.4)
$$u' = aite^{it} + ae^{it},$$
$$u'' = -ate^{it} + 2aie^{it},$$
$$u'' + u = 2aie^{it}.$$

Hence, the function (26.3) is a solution if $2ai = 1$, whence we see that

(26.5)
$$u = te^{it}/2i$$

is a particular solution. The general solution of Eq. (26.2) is thus of the form

(26.6)
$$u = c_1e^{it} + c_2e^{-it} + te^{it}/2i.$$

What does this mean? It tells us that the current in the system increases steadily in magnitude as t increases. Ultimately, this means disaster of one type or another, or at least a burnt-out fuse.

If the frequency of the voltage is different from the natural frequency of the system, we know that the current is the sum of two periodic currents. If, however, the frequency of the impressed voltage is the same as that of the natural frequency of the system, then the current will become arbitrarily large in magnitude over time. This phenomenon is called *resonance* and is of great theoretical and practical significance.

2.27 TUNING

As in the case of many physical phenomena, such as gravity, friction, etc., what is a nuisance in one situation may be an important asset in another. In the preceding section we considered the case where the frequency of the impressed voltage is varied and we studied the effect on the current in the system.

Let us now reverse our position and suppose that a fixed frequency of alternating voltage is being used and that our objective is to obtain as large a current as possible by adjusting the values of the parameters L, C, R. Suppose first that R is fixed, and that only L and C can be adjusted. It turns out, for technical reasons that we shall not go into, that it is easier to adjust C.

Referring to (24.6), we see that a critical value is again $\omega = 1/\sqrt{LC}$. The actual value to be used to maximize the amplitude of the current is a bit different, but it is clear that choosing $\omega = 1/\sqrt{LC}$ minimizes the denominator and thus helps increase the absolute value.

This process of varying the circuit parameters is called *tuning* and is precisely the operation engaged in when one uses a radio or television set. Here the impressed voltage is the signal sent out by the radio or TV station and picked up on the antenna. The actual circuit is considerably more complicated than that shown in Fig. 2.15, but the principle is the same.

2.28 UNIQUENESS: A REPRISE

We established the uniqueness of solution of

$$y'' + ay' + by = 0, \qquad y(0) = c_1, \qquad y'(0) = c_2, \tag{28.1}$$

by obtaining the solution explicitly. In what follows we shall carry out the same type of constructive demonstration for the equation

$$y'' + ay' + by = g(x), \qquad y(0) = c_1, \qquad y'(0) = c_2. \tag{28.2}$$

In studying more general equations and physical processes, we do not enjoy the luxury of an explicit analytic solution. Nonetheless, a uniqueness proof is essential. One way to proceed is the following. Let z be another solution of (28.2):

$$z'' + az' + bz = g(x), \qquad z(0) = c_1, \qquad z'(0) = c_2. \tag{28.3}$$

Subtracting Eq. (28.3) from Eq. (28.2), we see that

$$(y - z)'' + a(y - z)' + b(y - z) = 0. \tag{28.4}$$

Hence the function $u = y - z$ satisfies the homogeneous equation

$$u'' + au' + bu = 0, \tag{28.5}$$

with the initial conditions $u(0) = 0$, $u'(0) = 0$.

To establish the fact that u is identically zero for $x \geq 0$, we can proceed as follows. First, consider the special case $a = 0$, $b > 0$. If

$$u'' + bu = 0, \tag{28.6}$$

we have, upon multiplying by u' and integrating,

$$\int_0^x [u'(u'' + bu)]\, dx_1 = 0, \tag{28.7}$$

whence

$$\left[\frac{u'^2 + bu^2}{2}\right]_0^x = 0. \tag{28.8}$$

Since $u(0) = u'(0) = 0$, Eq. (28.8) yields

$$u'^2 + bu^2 = 0, \tag{28.9}$$

whence u and u' are identically zero, since $b > 0$. The reader may recognize this as a statement of the conservation of energy.

If $b < 0$, say $b = -1$, we obtain the equation

$$u'^2 - u^2 = 0. \tag{28.10}$$

It is no longer immediately clear that u is identically zero. We can factor this equation to obtain

$$(u' - u)(u' + u) = 0 \tag{28.11}$$

and carry through an argument based on the fact that Eq. (28.11) implies that either $u' - u = 0$ or $u' + u = 0$, but the proof is a bit artificial, and one which is neither instructive nor does it generalize to more general situations.

Let us instead pursue a different path, one of great importance in the general theory of differential equations.

2.29 UNIQUENESS: USE OF AN INTEGRAL EQUATION

To illustrate the method we wish to employ it is sufficient to consider the equation

$$u'' - u = 0, \qquad u(0) = 0, \qquad u'(0) = 0. \tag{29.1}$$

Integrating this equation once between 0 and x, we have

$$u'(x) = \int_0^x u(x_1)\, dx_1, \tag{29.2}$$

since $u'(0) = 0$. Integrating again, we have

$$u(x) = \int_0^x \left[\int_0^{x_1} u(x_2)\, dx_2\right] dx_1 = \int_0^x (x - x_1)u(x_1)\, dx_1, \tag{29.3}$$

as we see upon integration by parts. Hence

$$|u(x)| = \left|\int_0^x (x - x_1)u(x_1)\,dx_1\right| \le \int_0^x (x - x_1)|u(x_1)|\,dx_1. \tag{29.4}$$

Let us restrict our attention to a fixed interval $0 \le x \le a$. Then $(x - x_1) \le a$ and the preceding inequality leads to the further inequality

$$|u(x)| \le a \int_0^x |u(x_1)|\,dx_1, \tag{29.5}$$

and thus

$$|u(x)| \le b + a \int_0^x |u(x_1)|\,dx_1 \tag{29.6}$$

for any $b > 0$. Referring to Exercise 12 of Section 2.4, we see that (29.6) implies

$$|u(x)| \le be^{ax} \tag{29.7}$$

for $0 \le x \le a$. Since (29.7) holds for any $b > 0$, we must have $u(x) = 0$ for $0 \le x \le a$. Since a is arbitrary, u must be identically zero.

It is evident that we have used a certain amount of ingenuity and chicanery here and there. Unhappily, there is no way out of this in the cases where we do not possess general methods for the solution of the original equation. Our aim here has been to indicate to the reader how much can be done directly in terms of the original equation without ever using the explicit analytic form of the equation.

EXERCISES

1. If $a, b > 0$, we can proceed as follows. From

$$u'' + au' + bu = 0, \qquad u(0) = 0, \qquad u'(0) = 0,$$

we obtain upon multiplication by u' and integration between 0 and x, the equation $\frac{1}{2}u'^2 + \frac{1}{2}bu^2 + a\int_0^x u'^2\,dx_1 = 0$. Since all three terms are nonnegative, each must be zero, whence $u \equiv 0$.

2. Given that u satisfies

$$u'' + au' + bu = 0, \qquad u(0) = 0, \qquad u'(0) = 0,$$

show that we can always choose a positive constant k such that $v = e^{-kx}u$ satisfies an equation of the form

$$v'' + a_1v' + b_1v = 0, \qquad v(0) = 0, \qquad v'(0) = 0,$$

with $a_1, b_1 > 0$, regardless of the sign of a and b. Hence we have $v \equiv 0$ and $u \equiv 0$.

3. Show that

$$u(x) \leq a \int_0^x u(x_1)\, dx_1, \qquad u \geq 0,$$

implies that $u = 0$ by writing the first inequality in the form $dv/dx - av \leq 0$, where $v = \int_0^x u(x_1)\, dx_1$, and using the integrating factor e^{-ax}.

4. Carry through the argument for uniqueness of the solution $u = 0$ given in Section 2.29 for the equation

$$u'' + au' + bu = 0, \qquad u(0) = 0, \qquad u'(0) = 0.$$

5. Generalize the argument for uniqueness of the solution $u = 0$ given in Section 2.29 and apply it to the equation

$$u'' + a(x)u' + b(x)u = 0, \qquad u(0) = 0, \qquad u'(0) = 0,$$

under the assumption that $a(x)$ and $b(x)$ are continuous for $x \geq 0$.

6. Show that the change of variable $u = e^{-ax/2}v$ yields an equation for v which contains no term in v' if u satisfies $u'' + au' + bu = 0$.

7. Show that the change of variable

$$u = \exp\left[-\tfrac{1}{2} \int_0^x a(x_1)\, dx_1\right] v$$

yields an equation for v which contains no term in v' if u satisfies the equation

$$u'' + a(x)u' + b(x)u = 0.$$

2.30 GENERAL SOLUTION OF INHOMOGENEOUS EQUATION

Let us now apply the factorization techniques of Section 2.7 to obtain the general solution of

$$y'' + ay' + by = g(x). \tag{30.1}$$

Once we are in possession of this explicit representation, we have no further need for any special devices for obtaining particular solutions. Nonetheless, it is often quite useful, as we indicated in the previous sections, to use these special devices as analytical and algebraic shortcuts.

As before, let r_1 and r_2 denote the characteristic roots and write Eq. (30.1) in the form

(30.2) $$(y' - r_1 y)' - r_2(y' - r_1 y) = g(x).$$

Using the fact that (30.2) may be considered to be a first-order equation for the function $y' - r_1 y$, we obtain the relation

(30.3) $$y' - r_1 y = a_1 e^{r_2 x} + e^{r_2 x}\int_0^x e^{-r_2 x_1} g(x_1)\,dx_1,$$

where $a_1 = y'(0) - r_1 y(0)$. This is again a first-order equation, the general solution for which is

(30.4)
$$y = a_2 e^{r_1 x} + e^{r_1 x}\left\{\int_0^x e^{-r_1 x_1}\left[a_1 e^{r_2 x_1} + e^{r_2 x_1}\int_0^{x_1} e^{-r_2 x_2} g(x_2)\,dx_2\right]dx_1\right\}.$$

Let us separate the terms that are independent of g from the rest:

(30.5) $$y = a_2 e^{r_1 x} + a_1 e^{r_1 x}\int_0^x e^{-r_1 x_1}\,e^{r_2 x_1}\,dx_1$$
$$+ e^{r_1 x}\int_0^x e^{(r_2 - r_1)x_1}\left[\int_0^{x_1} e^{-r_2 x_2} g(x_2)\,dx_2\right]dx_1.$$

Upon carrying through the integration, we recognize the first part as the general solution of the homogeneous equation. Let us then concentrate on the second part. Integrating by parts, we have

(30.6)
$$e^{r_1 x}\left[\frac{e^{(r_2 - r_1)x}}{(r_2 - r_1)}\int_0^x e^{-r_2 x_2} g(x_2)\,dx_2\right]_0^x - e^{r_1 x}\int_0^x \frac{e^{(r_2 - r_1)x_1}}{(r_2 - r_1)}\,e^{-r_2 x_1} g(x_1)\,dx_1$$
$$= \frac{e^{r_2 x}}{(r_2 - r_1)}\int_0^x e^{-r_2 x_2} g(x_2)\,dx_2 - e^{r_1 x}\int_0^x \frac{e^{-r_1 x_1} g(x_1)}{(r_2 - r_1)}\,dx_1$$
$$= \int_0^x \left[\frac{e^{r_2(x - x_1)} - e^{r_1(x - x_1)}}{(r_2 - r_1)}\right] g(x_1)\,dx_1,$$

upon replacing the dummy variable x_2 by x_1 in the first integral. Formula (30.6) is a particular solution for Eq. (30.1). The general solution is obtained by adding a term of the form $c_1 e^{r_1 x} + c_2 e^{r_2 x}$.

EXERCISES

---

1. What is the form of the particular solution corresponding to (30.6) when the characteristic roots are equal?

2. Find the general solution of $u'' + au' + bu = t$. (Here $' = d/dt$.)
3. How many solutions are there of $u'' + 3u' + 2u = 1$ which are bounded as $t \to \infty$? How many of $u'' + u' - 2u = 1$? How many of $u'' - 3u' + 2u = 1$?
4. Show that if $b \neq 0$ there is always a solution of

$$u'' + au' + bu = p(t)$$

which is a polynomial of degree k if $p(t)$ is a polynomial of degree k. Find the solution for

$$u'' + u = t^2 - 2t + 2.$$

5. What is the solution of

$$u'' + u = f(t), \qquad u(0) = 0, \qquad u'(0) = 0,$$

where $f(t) = 1$, $0 \leq t \leq \pi$, $f(t) = 0$, $t > \pi$? Draw a graph of the solution for the interval $0 \leq t \leq 2\pi$.

2.31 TWO-POINT BOUNDARY CONDITIONS

Until now we have been considering linear differential equations whose solutions are determined by initial conditions. Let us discuss another and equally important way of specifying a particular solution. The simplified version of a neutron transport process described in Chapter 1 led to the system of linear differential equations

$$u'(x) = -pv(x), \qquad v'(x) = pu(x), \tag{31.1}$$

valid for $0 \leq x \leq a$, with the boundary conditions

$$u(a) = 0, \qquad v(0) = 1. \tag{31.2}$$

These conditions are quite different in kind from initial conditions, where u and v are both specified at either $x = 0$ or $x = a$. As we shall see, conditions of the type expressed by (31.2) produce effects distinct from those previously encountered. Indeed this is to be expected since the underlying physical processes are quite different.

There are several approaches to equations of this nature. Let us begin with the following method. Differentiating the first equation in (31.1), we have

$$u''(x) = -pv'(x). \tag{31.3}$$

Combining (31.3) with the second equation in (31.1), we obtain the equation

$$u'' = -p^2u, \qquad \text{or} \qquad u'' + p^2u = 0. \tag{31.4}$$

The conditions on u are

$$u(a) = 0, \qquad u'(0) = -p. \tag{31.5}$$

The first condition was given; the second is derived from the equation

$$u'(0) = -pv(0) = -p.$$

In the determination of u we begin with the fact that the general solution of (31.4) has the form

$$u = c_1 \cos px + c_2 \sin px, \tag{31.6}$$

where c_1 and c_2 are constants. Using the conditions (31.5), we obtain the two linear algebraic equations which determine c_1 and c_2:

$$0 = c_1 \cos pa + c_2 \sin pa, \qquad -p = pc_2. \tag{31.7}$$

Thus

$$c_2 = -1, \qquad c_1 = \frac{\sin pa}{\cos pa}. \tag{31.8}$$

Assume for the moment that $\cos pa \neq 0$. Using the above values of c_1 and c_2, we have

$$u = \frac{\sin pa}{\cos pa} \cos px - \sin px = \frac{\sin p(a - x)}{\cos pa}. \tag{31.9}$$

We thus obtain

$$u(0) = \frac{\sin pa}{\cos pa}, \tag{31.10}$$

$$v(a) = -\frac{u'(a)}{p} = \frac{1}{\cos pa}.$$

These values of u and v correspond respectively to the reflected and transmitted fluxes in terms of our one-dimensional transport process.

EXERCISES

1. Find the solution of

$$u'' + p^2u = 0, \qquad \int_0^a u\,dx = c_1, \qquad \int_0^a xu\,dx = c_2.$$

2. Solve

$$u'' + p^2u = 0, \qquad \int_0^a e^{b_1x}u\,dx = c_1, \qquad \int_0^a e^{b_2x}u\,dx = c_2.$$

3. Show without explicit use of the general solution that the solution of

$$u'' - p^2u = 0, \qquad u(a) = 0, \qquad u'(0) = 0,$$

is $u = 0$, by considering the integral $\int_0^a u'(u'' - p^2u)\,dt_1$.

4. Show that $u'' + u = 0$, $u(0) = u(\pi) = 0$, has infinitely many solutions.
5. Consider the solution of $u'' + b^2u = 0$, $u(0) = b_1$, $u(a) = b_2$. When does a unique solution exist?
6. Consider $u(0) = \tan pa$ as a function of a, $r(a)$, for $0 \le a < \pi/2$. Show directly that $r'(a) = p + pr(a)^2$, $r(0) = 0$.
7. Consider the equation $u'' - u = 0$, $u(0) = b_1$, $u(a) = b_2$. Show that there exists a unique solution regardless of the values of a, b_1, and b_2.
8. Consider the equation $u'' + u' + u = 0$, $u'(0) = -p$, $u(a) = 0$. Does there exist a unique solution for all $a > 0$? (The point we wish to emphasize is that two-point boundary-value problems are quite different from initial value problems. Initial value problems for linear equations with constant coefficients always possess a unique solution. Two-point boundary-value problems may not possess a solution at all, or may possess arbitrarily many solutions.)

2.32 CRITICAL LENGTH

The solution of Eqs. (31.4) and (31.5) has the form

$$u = \frac{\sin p(a - x)}{\cos pa} \tag{32.1}$$

for $0 \le x \le a < \pi/2p$. Here $\pi/2p$ enters as the first positive value of a for which $\cos pa = 0$. What interpretation do we give to the situation where $\cos pa = 0$? Begin with the case where a is small, so that $\cos pa$ is certainly nonzero. This corresponds to a short rod. As the rod increases in length, that is to say as a increases, there is more and more opportunity for neutron multiplication to proceed. Finally, when a reaches the value $\pi/2p$, we see that $u(x)$ is infinite for all values of x inside the rod. This value of a is the *critical length.*

2.33 THE EULER EQUATION

Linear differential equations with variable coefficients cannot in general be solved explicitly in terms of the elementary functions of analysis and a finite number of differentiations and integrations. Consequently, it is of interest to recognize various classes of such equations which can be transformed into equations with constant coefficients. One important class is the Euler equation

$$x^2y'' + axy' + by = 0. \tag{33.1}$$

Consider the effect of the change of independent variable,

$$x = e^t. \tag{33.2}$$

We have

$$\begin{aligned} \frac{dy}{dx} &= \frac{dt}{dx}\frac{dy}{dt} = e^{-t}\frac{dy}{dt}, \\ \frac{d^2y}{dx^2} &= \frac{d}{dx}\left(\frac{dy}{dx}\right) = \frac{d}{dx}\left(e^{-t}\frac{dy}{dt}\right) \\ &= \frac{dt}{dx}\frac{d}{dt}\left(e^{-t}\frac{dy}{dt}\right) \\ &= e^{-2t}\frac{d^2y}{dt^2} - e^{-2t}\frac{dy}{dt}. \end{aligned} \tag{33.3}$$

Hence, (33.1) becomes

$$\frac{d^2y}{dt^2} - \frac{dy}{dt} + a\frac{dy}{dt} + by = 0, \tag{33.4}$$

an equation with constant coefficients.

If we let r_1 and r_2 be the distinct roots of the quadratic equation

$$(r^2 - r) + ar + b = 0, \tag{33.5}$$

then the general solution of (33.4) has the form

$$y = c_1e^{r_1t} + c_2e^{r_2t}. \tag{33.6}$$

Hence, recalling (33.2), the general solution of (33.1) has the form

$$y = c_1x^{r_1} + c_2x^{r_2}. \tag{33.7}$$

EXERCISES

1. Obtain the solution of $x^2y'' + xy' = 0$, $y(1) = 1$, $y'(1) = 2$.
2. Obtain the solution of $x^2y'' - 2y = 0$, $y(1) = 0$, $y'(1) = 1$.

2.34 LINEAR DIFFERENCE EQUATIONS

In a number of physical processes it is convenient to observe the variables of importance only at the discrete times, $t = 0, 1, 2, \ldots$, or to employ only the values at these times in a mathematical analysis. In place of a differential equation of the type we have been considering, we can obtain in this fashion a *difference equation*,

$$u_{n+2} + au_{n+1} + bu_n = g_n, \tag{34.1}$$

$n = 0, 1, 2, \ldots$, with $u_0 = c_1$, $u_1 = c_2$. In place of writing $u(n)$, it is customary to use the subscript notation u_n.

The theory of equations of this nature is parallel in many ways to that for differential equations, and, generally, easier as far as existence and uniqueness of solutions is concerned. Difference equations are of fundamental importance in connection with the numerical solution of differential equations. As we shall see in Chapter 4 a fundamental technique in the computational solution of differential equations is to transform the original differential equation into an approximating difference equation.

Let us consider first the homogeneous equation

$$u_{n+2} + au_{n+1} + bu_n = 0, \qquad n \geq 0, \quad u_0 = c_1, \quad u_1 = c_2. \tag{34.2}$$

Again c_1 and c_2 are the *initial conditions* for the equation. The equation determines u_2 in terms of u_1 and u_0, u_3 in terms of u_2 and u_1, and so on. Thus, if u_0 and u_1 are specified, the value u_n is uniquely determined for every n by these values. The problem remains to derive a convenient analytic representation. To do this, we imitate the procedure for differential equations.

Let us look for a particular solution of the form r^n. Substituting in the difference equation, we have

$$r^{n+2} + ar^{n+1} + br^n = r^n(r^2 + ar + b) = 0. \tag{34.3}$$

Hence, if r is a root of the *characteristic equation*

$$r^2 + ar + b = 0, \tag{34.4}$$

then r^n is a solution. Assume for the moment that the roots are distinct and let them be designated by r_1 and r_2. Then

$$u_n = a_1r_1^n + a_2r_2^n \tag{34.5}$$

is a solution of (34.2), apart from the initial conditions. Let us show that it is the general solution by showing that we can choose a_1 and a_2 so as to satisfy any initial conditions of the form

$$u_0 = c_1, \qquad u_1 = c_2. \tag{34.6}$$

Using (34.5), we have

$$\begin{aligned} c_1 &= a_1 + a_2, \\ c_2 &= a_1 r_1 + a_2 r_2. \end{aligned} \tag{34.7}$$

The determinant of the coefficients is

$$\begin{vmatrix} 1 & 1 \\ r_1 & r_2 \end{vmatrix} = r_2 - r_1 \neq 0, \tag{34.8}$$

since $r_2 \neq r_1$ by assumption. Hence a_1 and a_2 can be obtained from (34.7).

If the roots of the characteristic equation are equal, we guess by analogy with the corresponding result for differential equations that nr_1^n is a second solution and, thus, that the general solution of (34.2) now takes the form

$$u_n = a_1 r_1^n + a_2 n r_1^n. \tag{34.9}$$

To fit the initial conditions, $u_0 = c_1$, $u_1 = c_2$, we have the equations

$$\begin{aligned} c_1 &= a_1, \\ c_2 &= a_1 r_1 + a_2 r_1. \end{aligned} \tag{34.10}$$

The determinant is now

$$\begin{vmatrix} 1 & 0 \\ r_1 & r_1 \end{vmatrix} = r_1, \tag{34.11}$$

which is nonzero if $r_1 \neq 0$. Hence, (34.9) is the general solution for $n = 0, 1, 2, \ldots$ provided that the double root is not $r_1 = 0$ (which occurs only if $a = b = 0$).

EXERCISES

1. Consider the first-order difference equation $u_{n+1} = au_n$, $u_0 = c$. Show that $u_n = ca^n$, $n = 0, 1, \ldots$.
2. Consider the *Fibonacci sequence* $\{u_n\}$ defined by $u_{n+2} = u_{n+1} + u_n$, $u_0 = 2$, $u_1 = 1$. Show that

$$u_n = \left(\frac{1 + \sqrt{5}}{2}\right)^n + \left(\frac{1 - \sqrt{5}}{2}\right)^n.$$

3. What is the limiting ratio of u_{n+1}/u_n as $n \to \infty$?
4. Consider the linear difference equation $u_{n+2} - u_n = 0, u_0 = 1, u_1 = 1$. Show that $u_n = 1$ for $n \geq 1$.
5. With the same equation and the initial conditions $u_0 = 1$, $u_1 = 0$, show that $u_n = [1 + (-1)^n]/2$.
6. What is a necessary and sufficient condition on the coefficients a and b that all solutions of $u_{n+2} + au_{n+1} + bu_n = 0$ approach zero as $n \to \infty$?
7. Find a particular solution of $u_{n+2} + au_{n+1} + bu_n = s^n$ if s is not a characteristic root. What is a particular solution if s is a characteristic root?
8. Find the general solution of $u_{n+2} - 2u_{n+1} + 2u_n = 0$, and find the particular solution for which $u_0 = 1$, $u_1 = 1$. Show that the latter can be written $u_n = 2^{n/2} \cos (n\pi/4)$.
9. Find the solution of $u_{n+2} = u_{n+1} + u_n + 2$, $u_0 = 1$, $u_1 = 1$.
10. Show that the solution of $u_{n+1} = a_n u_n$, $u_0 = c$ is given by

$$u_n = \left(\prod_{k=0}^{n-1} a_k\right) c.$$

11. Find the solution of $u_{n+1} = a_n u_n + f_n$, $u_0 = c$. [*Hint:* Set $u_n = (\prod_{k=0}^{n-1} a_k)v_n$ and consider the equation for v_n.]
12. Write the solution of $u_{n+1} = au_n + f_n$, $u_0 = c$.
13. What is the limiting form of u_n as $n \to \infty$, where $u_{n+1} = 1 + u_n/2$, $u_0 = 1$?
14. Show that $u_{n+2} + au_{n+1} + bu_n = 0$, $u_0 = c_1$, $u_N = c_2$, has a unique solution if $r_1^N \neq r_2^N$, where r_1 and r_2 are the characteristic roots.
15. Obtain the solution of $u_{n+1} - u_{n-1} = 2\,\Delta u_n$, $n = 1, 2, \ldots$, with $u_0 = 1, u_1 = e^{-\Delta}$. Compare the value of u_n with that of $v_{n+1} - v_{n-1} = 2\,\Delta v_n$, $n = 1, 2, \ldots, v_0 = 1, v_1 = 1 - \Delta$, and $v_1 = 1 - \Delta + \Delta^2/2$.
16. Compare u_n of Ex. 15 with the solution of $w_n - w_{n-1} = \Delta w_n$, $n = 1, 2, \ldots, w_0 = 1$.
17. (The Gambler's ruin problem). Two men are tossing fair coins. One wins if the coins match, the other wins if not. The first has p coins initially, the other q. What is the probability that the first player eventually wins all the coins of the other? (Consider a general situation in the course of play where the first player has n coins, $0 < n < p + q$. Let u_n be the probability that the first player eventually wins all of the other player's coins. Then $u_n = \frac{1}{2}(u_{n-1} + u_{n+1})$, $0 < n < p + q$, and $u_0 = 0$, $u_{p+q} = 1$.)

18. If $u_{n+2} = au_{n+1} + bu_n$, $u_0 = 1$, $u_1 = c$, show that

$$\sum_{n=0}^{\infty} u_n r^n = \frac{1 + r(c - a)}{1 - r(a + br)}.$$

[*Hint:* Write

$$f(r) = \sum_{n=0}^{\infty} u_n r^n = 1 + cr + \sum_{n=0}^{\infty} u_{n+2} r^{n+2}$$

$$= 1 + cr + r^2 \sum_{n=0}^{\infty} [au_{n+1} + bu_n] r^n,$$

and thus obtain an equation for $f(r)$. The function $f(r)$ is called the *generating function* of the sequence $\{u_n\}$. The generating function is occasionally called the "z-transform" in engineering circles. A continuous version is the *Laplace transform* we briefly discuss in Chapter 5.]

19. If $u_{n+2} = 3u_{n+1} - 2u_n$, $u_0 = 1$, $u_1 = 0$, show that

$$\sum_{n=0}^{\infty} u_n r^n = \frac{1 - 3r}{1 - 3r + 2r^2} = \frac{2}{1 - r} - \frac{1}{1 - 2r},$$

and thus show that $u_n = 2 - 2^n$, $n = 0, 1, \ldots$. [*Hint:* $2/(1 - r) = 2 + 2r + 2r^2 + \cdots$; $1/1 - 2r = 1 + 2r + 2^2r^2 + \cdots$.]

2.35 DISCUSSION

In this chapter we have considered various types of solutions of the linear differential equation

$$u'' + au' + bu = g(x), \tag{35.1}$$

and the behavior of the solutions as a function of the independent variable, as a function of the coefficients, and as a function of the length of the interval over which the solution is determined.

What we have shown is that a number of fundamental phenomena, such as periodicity, damping, natural frequency, resonance, chain reaction, can be explained to a greater or lesser extent in terms of the foregoing simple equation. If we want better explanations and explanations of more complex phenomena, then we must consider more complicated equations. This is a chore the mathematician accepts with pleasure.

MISCELLANEOUS EXERCISES

1. Obtain the general solution of the equation

$$y'' + ay' + by = g(x)$$

in the following fashion.

a) Let y_1 and y_2 be two independent solutions of the homogeneous differential equation $y'' + ay' + by = 0$. Write $y = u_1y_1 + u_2y_2$, where u_1 and u_2 are functions to be determined in an adroit fashion.

b) Taking derivatives, we have

$$y' = u_1y_1' + u_2y_2' + u_1'y_1 + u_2'y_2.$$

Set $u_1'y_1' + u_2'y_2 = 0$.

c) Using $y' = u_1y_1' + u_2y_2'$, calculate y'' and then set

$$y'' + ay' + by = g(x).$$

In this way, obtain a second linear algebraic equation for u_1' and u_2'.

d) Using the two linear algebraic equations obtained from (b) and (c), solve for u_1' and u_2'. From this solution obtain u_1 and u_2 and thus the general solution of the original inhomogeneous equation. (This procedure is a special case of the famous technique of "variation of parameters" due to Lagrange.)

2. Consider the equation

$$\epsilon u'' + (1 + \epsilon)u' + u = 0, \qquad u(0) = 1, \qquad u'(0) = 0,$$

where $\epsilon > 0$. Write the solution in the form $u(t, \epsilon)$. Is it true that $u(t,\epsilon)$ is a continuous function of ϵ for $\epsilon > 0$ and a fixed $t \geq 0$? Does $\lim_{\epsilon \to 0} u(t, \epsilon)$ exist for any $t \geq 0$? What is the connection, if any, between this limit function and the solution of $u' + u = 0$, $u(0) = 1$?

3. Consider similar questions for

$$u'' + (1 + \epsilon)u = 0, \qquad u'' + (1 + \epsilon)u' + u = 0.$$

(The point of the foregoing examples is that sometimes one can use the limiting equation with impunity and sometimes not. Problems of this nature are never routine; each must be carefully investigated.)

4. Obtain the solution of

$$u = 1 + \int_0^1 e^{k|x-x_1|} u(x_1)\, dx_1$$

by reducing the equation to a second-order differential equation subject to a two-point boundary condition. [*Hint:* Write

$$\int_0^1 e^{k|x-x_1|} u(x_1)\, dx_1 = \int_0^x e^{k(x-x_1)} u(x_1)\, dx_1$$

$$+ \int_x^1 e^{-k(x-x_1)} u(x_1)\, dx_1].$$

5. Obtain the solution of

$$u = f(x) + \int_0^1 e^{k|x-x_1|} u(x_1)\, dx_1,$$

under the assumption that $f(x)$ possesses derivatives.

6. Show that the maximum of the function $f(v) = 2uv - v^2$ taken over all v is u^2.

7. Let $g(u)$ be a function of u satisfying the condition $g''(u) > 0$ for all u. Show that the maximum of $f(v) = g(v) + (u - v)g'(v)$ over all v is $g(u)$. Thus, we can write

$$g(u) = \max_v\, [g(v) + (u - v)g'(v)].$$

8. What is the geometric significance of the result of Exercise 7? [*Hint:* Consider the location of the tangent to the curve $w = g(u)$.]

9. Consider the equation

$$u' = u^2 + 1$$

and the companion equation

$$w' = 2vw - v^2 + 1,$$

both subject to the same initial condition $u(0) = w(0) = 1$, in the interval $0 \leq t < \pi/2$, where v is any continuous function of t in $0 \leq t < \pi/2$. Show that $w(t) \leq u(t)$ in this interval regardless of the choice of $v(t)$. [*Hint:* Use the fact that $2vw - v^2 \leq w^2$ for all v and w.]

10. Write the explicit solution of the equation $w' = 2vw - v^2 + 1$, $w(0) = c$, in the form

$$\begin{aligned} w &= c \exp\left(2 \int_0^t v \, dt_1\right) + \exp\left(2 \int_0^t v \, dt_1\right) \\ &\quad \times \left[\int_0^t (1 - v^2) \exp\left(-2 \int_0^{t_1} v \, dt_2\right) dt_1\right] \\ &= T(v, t), \end{aligned}$$

and show that $u(t) = \max_v T(v, t)$ for $0 \leq t < \pi/2$.

11. Obtain the solution of $y' = |y - 1|$, $y(0) = a$. [*Hint:* Consider first the case where $a > 1$. Then in the neighborhood of $x = 0$, the equation has the form $y' = y - 1$, $y(0) = a$. The solution is $y = 1 + (a - 1)e^t$. Hence $y > 1$ for $t \geq 0$, and the solution is as given. Consider next the case where $a < 1$. Then in the neighborhood of $x = 0$, $y' = 1 - y$, $y(0) = a$ with the solution $y = 1 + (a - 1)e^{-t}$. Hence $y < 1$ for $t \geq 0$ and the solution is as given. If $a = 1$, then $y = 1$ for $t \geq 0$.]

12. Discuss similarly the equation $u' = |(1 - u)(2 - u)|$, $u(0) = a$.

13. Consider the equation $y'' + 3|y'| + 2y = 0$, $y(0) = c_1$, $y'(0) = c_2$. [*Hint:* Consider first the case where $y'(0) > 0$ and then where $y'(0) < 0$. Equations of this nature arise frequently in modern control theory, particularly in connection with "bang-bang" control.]

14. In Exercise 1, at step (b), set $u_1'y_1 + u_2'y_2 = f(x)$, where $f(x)$ is any differentiable function of x. Now proceed as explained in steps (c) and (d) to obtain the general solution of $y'' + ay' + by = g(x)$.

BIBLIOGRAPHY AND COMMENTS

Section 2.4. The inequality in Exercise 12 is one of the most useful tools in the theory of differential equations. For some applications, see:

BELLMAN, R., *Stability Theory of Differential Equations*, Dover Publications, New York, 1969.

Section 2.6. Discontinuous forcing terms arise when an external force is applied to a system starting at time $t = t_0$. Discontinuous coefficients arise in the study of particle or wave propagation through stratified, or layered, media.

Section 2.13. The Riccati equation plays a major role in modern control theory and in mathematical physics. See:

BELLMAN, R., *Introduction to the Mathematical Theory of Control Processes*, Vol. I, Academic Press, Inc., New York, 1968.

Section 2.14. The idea of representing the solution of a differential equation as a curve $y = y(x)$ traversed over time is quite natural in view of the origins of differential equations. The idea of considering (y, y') as the coordinates of a point in the phase plane is a bit more ingenious and has applications of great significance in the general theory of nonlinear second-order differential equations. See:

LEFSCHETZ, S., *Differential Equations, Geometric Theory*, Interscience, New York, 1957.

Section 2.15. In Exercise 2 of this section and in the exercises at the end of Section 2.13 we exposed the reader to one of the most interesting of scientific problems, the identification problem. Given observations of a system at various times, can we determine the structure of the system? For some discussions of this question, see:

BELLMAN, R., and R. KALABA, *Quasilinearization and Nonlinear Boundary Value Problems*, American Elsevier, New York, 1965.

Section 2.29. What we were hinting at here was the surprising fact that the standard way to obtain important results in the theory of differential equations is to convert the equation into a suitably chosen integral equation.

CHAPTER 3

Power-Series Solutions

3.1 INTRODUCTION

In Chapter 1, we illustrated some of the ways in which the study of scientific phenomena gave rise to differential equations. In Chapter 2, we considered a particular class of equations—linear equations with constant coefficients—and used the explicit analytic solution to discuss some important processes.

In this chapter we wish to present a different and more general technique for obtaining solutions of differential equations: the method of solution in terms of power series. This approach is a general factotum of mathematical analysis, extensively employed in all realms of science where quantitative methods can be used. The simple idea, once grasped, can be applied in a surprising number of ways. We shall demonstrate how this method can be used to obtain numerical solutions of both linear and nonlinear equations. In particular, it includes the fundamental technique of perturbation analysis. We shall give a number of simple applications, and point out the types of problems which require a more profound analysis.

The idea of a power-series expansion will be of great use to us in the following chapter where we discuss the numerical solution of differential equations using a digital computer.

3.2 POLYNOMIALS AND POWER SERIES

In elementary algebra, we study functions of the form

$$a_0 + a_1x + a_2x^2 + \cdots + a_nx^n, \tag{2.1}$$

the polynomials, because they behave so simply and sensibly under the operations of arithmetic. The sum of two polynomials is again a polynomial,

$$(2.2)\qquad (a_0 + a_1x + \cdots + a_nx^n) + (b_0 + b_1x + \cdots + b_nx^n) = (a_0 + b_0) + (a_1 + b_1)x + \cdots + (a_n + b_n)x^n;$$

and likewise the product of two polynomials is a polynomial,

$$(2.3)\qquad (a_0 + a_1x + \cdots + a_nx^n)(b_0 + b_1x + \cdots + b_nx^n) = a_0b_0 + (a_1b_0 + a_0b_1)x + \cdots + a_nb_nx^{2n}.$$

The importance of these functions in calculus derives from the fundamental formula

$$(2.4)\qquad \frac{d}{dx}x^n = nx^{n-1}$$

for any exponent n. Hence the derivative of a polynomial is a polynomial,

$$(2.5)\qquad \frac{d}{dt}(a_0 + a_1x + \cdots + a_nx^n) = a_1 + 2a_2x + \cdots + na_nx^{n-1};$$

and similarly the indefinite integral of a polynomial is a polynomial.

In calculus, where the concept of limit is added to the basic operations of algebra, we are naturally led to study, as an extension of polynomials, the class of power series:

$$(2.6)\qquad a_0 + a_1x + a_2x^2 + \cdots + a_nx^n + \cdots$$

This is the limiting form of a polynomial as the degree increases indefinitely. The introduction of power series enables us to consider the division of polynomials. For, if $a_0 \neq 0$, the reciprocal of a power series is also a power series:

$$(2.7)\qquad (a_0 + a_1x + \cdots + a_nx^n + \cdots)^{-1} = a_0^{-1} - a_1a_0^{-2}x - \cdots$$

Power series, however, are very much more complex than polynomials because of the problem of *convergence*, a question we shall review in the following section. Leaving this aside for the moment, let us point out that the properties stated above allow power series to qualify as candidates for solutions of differential equations such as

$$(2.8a)\qquad u'' + (1 + x)u = 0$$

or

$$(2.8b)\qquad xu'' + u' + (n^2 - x^2)u = 0.$$

And, indeed, this is the reason for the apparent digression from the domain of differential equations.

3.3 CONVERGENCE AND DIVERGENCE

We shall suppose that the reader has had an introductory course in calculus, or has access to a book on the subject, and preferably both. Consequently, we wish merely to review some familiar facts concerning the convergence of power series. For our purposes it is sufficient to suppose that variable x is real, and takes values in some interval $(-c, c)$.

As we see from the familiar examples,

$$1 + x + x^2 + \cdots + x^n + \cdots, \tag{3.1a}$$

$$1 + x + \frac{x^2}{2!} + \cdots + \frac{x^n}{n!} + \cdots, \tag{3.1b}$$

$$1 + x + 2!x + \cdots + n!x^n + \cdots, \tag{3.1c}$$

a power series may converge somewhere: in a symmetric interval about $x = 0$ [$|x| < 1$ in the case of (3.1a)]; everywhere as in the case of (3.1b); or only for $x = 0$, as with (3.1c). The reader must not think, however, that thoroughly divergent series are of no significance in either theory or application. Under the name "asymptotic series" they enter in a most important way in the advanced theory of differential equations. However, we shall not consider this interesting topic here.

The standard technique for establishing convergence is by comparison with a series with known properties, such as (3.1a), (3.1b), or (3.1c). Since we shall use any results of this type in a very sparing fashion, we shall refer the reader to his calculus book if he wishes to recall results of this nature.

Let us recall some important properties concerning power series which we will employ in the ensuing discussion. Let $f(x)$ and $g(x)$ be given, respectively, in terms of the power series

$$f(x) = a_0 + a_1x + a_2x^2 + \cdots + a_nx^n + \cdots, \tag{3.2a}$$

$$g(x) = b_0 + b_1x + b_2x^2 + \cdots + b_nx^n + \cdots, \tag{3.2b}$$

both convergent for $|x| < 1$.

Then we can add, subtract, and multiply the expressions for $f(x)$ and $g(x)$ as if they were polynomials and obtain power series which converge in the same interval. Thus

(3.3a)

$$f(x) + g(x) = (a_0 + b_0) + (a_1 + b_1)x + \cdots + (a_n + b_n)x^n + \cdots,$$

(3.3b)

$$f(x)g(x) = a_0b_0 + (a_0b_1 + a_1b_0)x + (a_2b_0 + a_1b_1 + a_0b_2)x^2 + \cdots$$

Both of these new series are convergent for $|x| < 1$.

EXERCISES

1. Show that
$$f^2(x) = a_0^2 + 2a_0a_1x + (a_1^2 + 2a_0a_2)x^2 + \cdots,$$
and write the expression for the general term, where $f(x)$ is given in (3.2a).
2. Obtain the first four terms in the power-series expansions of e^f and $\sin f$, where f is as given in (3.2a).
3. Obtain the first four terms in the power-series expansion of $f(g(x))$, where f and g are as given in (3.2a) and (3.2b).

3.4 MACLAURIN SERIES

Suppose we are told that a function $f(x)$, defined say for $|x| < 1$, possesses a power-series expansion in the given interval:

$$f(x) = a_0 + a_1x + \cdots + a_nx^n + \cdots \tag{4.1}$$

How do we determine the coefficients?

The determination of the first one, a_0, is simple. Setting $x = 0$, we have

$$f(0) = a_0. \tag{4.2}$$

To obtain a_1, we must invoke the following theorem:

Theorem. *If $f(x)$ possesses the power-series expansion* (4.1) *in the interval $|x| < 1$, then $f'(x)$ possesses the power-series expansion*

$$f'(x) = a_1 + 2a_2x + \cdots + na_nx^{n-1} + \cdots \tag{4.3}$$

in the same interval.

Observe that the right-hand side is exactly what we obtain by differentiating the power series (4.1) term by term. The proof for this theorem is not difficult, but consistent with our previous policy, we refer the reader to any standard text on calculus.

Using the power-series expansion (4.3) and once again setting $x = 0$, we obtain

$$f'(0) = a_1. \tag{4.4}$$

Using the theorem once again, we have

$$f''(x) = 2a_2 + 6a_3x + \cdots, \tag{4.5}$$

whence, as above,

$$\frac{f''(0)}{2} = a_2. \tag{4.6}$$

Continuing in this way, we obtain the familiar formula

$$a_n = \frac{f^{(n)}(0)}{n!}, \qquad n = 1, 2, \ldots, \tag{4.7}$$

for the coefficient of x^n in the power-series expansion of $f(x)$. The series obtained in this way is called the Maclaurin series.

EXERCISES

1. Using the method outlined in this section, obtain the power-series expansion of e^x, $\cos x$, and $\sin x$.
2. Obtain the power-series expansion of $(1 - x)^a$. Show that the series terminates if and only if a is a positive integer.
3. Show that if $a_0 + a_1x + \cdots + a_nx^n + \cdots$ converges in the interval $|x| < 1$ and is identically zero in this interval, then all coefficients are zero.
4. Prove Exercise 3 *without* using differentiation. [*Hint:* Prove $a_0 = 0$. Then show that $a_1 = 0$, and use induction.]
5. Show that if

   $$f(x) = b_0 + b_1(x - a) + b_2(x - a)^2 + \cdots$$

   in an interval about $x = a$, then

   $$b_n = \frac{f^{(n)}(a)}{n!}, \qquad n = 1, 2, \ldots,$$

   with $b_0 = a_0$. [The formula holds for $n = 0$ with a convenient definition of $f^{(0)}(a)$ and $0!$.]
6. Use the power series

   $$\frac{1}{(1 - x)} = 1 + x + x^2 + \cdots + x^n + \cdots,$$

   a geometric series, to deduce that

   $$\log(1 - x) = -x - \frac{x^2}{2} - \cdots - \frac{x^n}{n} \cdots$$

7. Show that

$$\log \frac{1+x}{1-x} = 2x + \frac{2x^3}{3} + \frac{2x^5}{5} + \cdots$$

8. Let a be an approximate solution of $f(x) = 0$. Expand $f(x)$ in the neighborhood of $x = a$,

$$f(x) = f(a) + (x - a)f'(a) + \cdots,$$

and retain the first two terms. Show that $x = a - f(a)/f'(a)$ is an improved approximation, if a is indeed close to the actual solution. What is the graphical interpretation of this result? How do we obtain a still better approximation? (This is called the Newton-Raphson approximation method.)

9. Consider the equation $1 - x + \frac{1}{10}x^3 = 0$. It has an approximate root $x_1 = 1$. Obtain two further improvements using the technique given in Exercise 8.

10. Similarly, obtain an approximate value for $\sqrt{2}$ using the equation $x^2 - 2 = 0$.

3.5 POWER-SERIES SOLUTIONS OF DIFFERENTIAL EQUATIONS

Our purpose in reviewing and recalling these basic results concerning power series has been to set the stage for the use of power series as solutions of differential equations. Our emphasis is on the methodology, since rigorous derivations require a more advanced level than we wish to invoke here. In all cases, we either state what can be proved, or freely admit that we are using the method in the hope that it works.

The importance of examining the convergence of the series that are obtained as a result of using power-series expansions in connection with differential equations can scarcely be overemphasized. Numerous examples exist of the fallacious results that can be obtained by means of a formal application of expansion techniques.

This is particularly significant if numerical solutions are desired, with or without the use of a digital computer. A preliminary study of convergence, together with an estimate of the error committed in using the partial sum

$$S_n(x) = a_0 + a_1x + \cdots + a_nx^n \tag{5.1}$$

as an approximation to the value of the infinite series

$$S(x) = a_0 + a_1x + \cdots + a_nx^n + \cdots, \tag{5.2}$$

is essential in determining both the amount of effort involved in the numerical calculation and the kind of effort that should be used.

Even if convergence has been established, there remain the far from trivial questions concerning the degree of accuracy that can be obtained with a reasonable amount of effort. Thus the series

$$\log 2 = 1 - \tfrac{1}{2} + \tfrac{1}{3} - \cdots, \tag{5.3a}$$

$$\pi/4 = 1 - \tfrac{1}{3} + \tfrac{1}{5} - \cdots, \tag{5.3b}$$

are both convergent but hardly suitable for calculating either of these fundamental constants to ten or twenty significant figures.

When a preliminary analysis of convergence is beyond our capabilities, there remains the experimental approach of calculating $S_n(x)$ for $n = 0, 1, 2, \ldots$, and seeing whether or not convergence occurs. In a number of important cases, the series may diverge, and yet it may be true that a particular partial sum yields an excellent approximation to the desired solution. This is the basic idea of *asymptotic series.*

3.6 EXISTENCE AND UNIQUENESS OF SOLUTION

To illustrate the use of power series, we consider the equation

$$u'' - xu = 0, \qquad u(0) = 1, \qquad u'(0) = 0. \tag{6.1}$$

The first question we must ask ourselves is: Does there exist *a* solution of this equation? If the answer is affirmative, the second question is: How many solutions are there? These are *existence* and *uniqueness* problems, matters of extreme importance from both the theoretical and applied points of view.

Before attempting to determine a power-series expansion for a solution, it is clear that we wish to know that one exists. If more than one exists, we naturally want to know which one we are finding. From the standpoint of constructing a mathematical model of a physical process, a lack of existence shows that we have demanded too much in our equations and a lack of uniqueness shows that we have not demanded enough. The art of mathematical model-making consists of finding a suitable balance between the demands of science and our ability to handle complex equations.

In the next chapter, as a preliminary to numerical solution, we go into this subject in some further detail. At the moment, we ask the reader to proceed on faith and to accept the fact that each equation he sees in this chapter has a unique solution.

The existence of a solution is actually not hard to demonstrate. We shall exhibit a power series which, when substituted into Eq. (6.1), satisfies the equation identically. The uniqueness of the power-series solution follows immediately from our approach. Uniqueness of solution, in general, requires a bit more which we do not think essential in a first course. We shall, however, present some fundamental results in this area in Chapter 6.

3.7 POWER-SERIES SOLUTION OF $u'' - xu = 0$

Let us now see whether we can obtain a power-series solution to the equation

$$u'' - xu = 0, \qquad u(0) = 1, \qquad u'(0) = 0. \tag{7.1}$$

Let us first present the method, which is quite simple. Once we have demonstrated that the technique is applicable, we have some motivation for discussing convergence and divergence of the power series we obtain.

Assume that there is a solution of (7.1) of the form

$$u = a_0 + a_1x + a_2x^2 + \cdots + a_nx^n + \cdots, \tag{7.2}$$

convergent in some interval $|x| < b$. For the moment, we ignore the initial conditions. Then, as we mentioned in Section 3.4, term-by-term differentiation is legitimate in this interval, and we have

$$\begin{aligned} u' &= a_1 + 2a_2x + \cdots + na_nx^{n-1} + \cdots, \\ u'' &= 2a_2 + 6a_3x + \cdots + n(n-1)a_nx^{n-2} + \cdots \end{aligned} \tag{7.3}$$

Thus

$$\begin{aligned} u'' - xu = 2a_2 &+ (6a_3 - a_0)x + (12a_4 - a_1)x^2 \\ &+ (20a_5 - a_2)x^3 + \cdots \\ &+ [(n+2)(n+1)a_{n+2} - a_{n-1}]x^n + \cdots \end{aligned} \tag{7.4}$$

If the right-hand side is to be identically zero, in order that Eq. (7.1) hold, all of the coefficients in the power series must be zero (see Exercise 3 of Section 3.4):

$$\begin{aligned} 2a_2 &= 0, & 6a_3 - a_0 &= 0, \\ 12a_4 - a_1 &= 0, & 20a_5 - a_2 &= 0, \end{aligned} \tag{7.5}$$

and, generally,

$$(n+2)(n+1)a_{n+2} - a_{n-1} = 0, \tag{7.6}$$

for $n = 1, 2, \ldots$ We see that

(7.7)
$$a_3 = \frac{a_0}{6}, \qquad a_6 = \frac{a_3}{6 \cdot 5} = \frac{a_0}{6 \cdot 5 \cdot 3 \cdot 2}, \ldots,$$
$$a_2 = 0, \qquad a_5 = \frac{a_2}{5 \cdot 4} = 0, \ldots,$$
$$a_4 = \frac{a_1}{4 \cdot 3}, \qquad a_7 = \frac{a_4}{7 \cdot 6} = \frac{a_1}{7 \cdot 6 \cdot 4 \cdot 3}, \ldots$$

Let us now use the initial conditions, $u(0) = 1$, $u'(0) = 0$. We have then $a_0 = 1$, $a_1 = 0$.

Hence only $a_0, a_3, a_6, \ldots, a_{3n}, \ldots,$ are distinct from zero. Using the relations (7.7), we are led to the series

(7.8)
$$u(x) = 1 + \frac{x^3}{3 \cdot 2} + \frac{x^6}{6 \cdot 5 \cdot 3 \cdot 2} + \cdots + \frac{x^{3n}}{3n(3n-1)(3n-3)(3n-4)} + \cdots.$$

Similarly, we find that the solution determined by $u(0) = 0$, $u'(0) = 1$ is given by

(7.9)
$$u(x) = x + \frac{x^4}{4 \cdot 3} + \frac{x^7}{7 \cdot 6 \cdot 4 \cdot 3} + \cdots$$

The solutions u_1 and u_2, together with the initial conditions $u_1(0) = 1$, $u_1'(0) = 0$, $u_2(0) = 0$, $u_2'(0) = 1$, are called the *principal solutions.*

The power-series solution subject to the conditions $u(0) = c_1$, $u'(0) = c_2$ is given by $u = c_1u_1 + c_2u_2$.

EXERCISES

1. Find a power series solution of each of the following initial-value problems. If possible find the general term of the series.
   a) $u'' - u = 0$, $\quad u(0) = 1,\ u'(0) = 6$
   b) $u'' - xu' - u = 0$, $\quad u(0) = 2,\ u'(0) = 0$
   c) $u'' + 4x^2u = 0$, $\quad u(0) = 0,\ u'(0) = 1$
   d) $u'' - 2xu' + 2u = 0$, $\quad u(0) = 0,\ u'(0) = 1.$

2. Show that the principal solutions of $u'' - x^2u = 0$ start out as follows:
$$u_1(x) = 1 + \frac{x^4}{4 \cdot 3} + \cdots, \qquad u_2(x) = x + \frac{x^5}{5 \cdot 4} + \cdots$$
   Find the general terms.

3. Show that the principal solutions of $u'' - x^k u = 0$, for $k = 1, 2, \ldots$, start out as follows:

$$u_1(x) = 1 + \frac{x^{k+2}}{(k+2)(k+1)} + \cdots,$$

$$u_2(x) = x + \frac{x^{k+3}}{(k+3)(k+2)} + \cdots$$

Find the general terms.

4. Show that if we start with $u' - xu = 0$, $u(0) = 1$, we obtain the power-series expansion for $e^{x^2/2}$.
5. Consider the equation $u' - 2xu = 1$, $u(0) = 0$. Obtain a power-series solution having the form $u = x + a_2 x^2 + \cdots$, and in this way obtain a power series expansion for the function

$$e^{x^2} \int_0^x e^{-x_1^2}\, dx_1.$$

6. Show that the function $(\text{arc sin } x)^2$ satisfies the equation

$$(1 - x^2)u'' - xu' = 2, \qquad u(0) = 0, \qquad u'(0) = 0,$$

and in this way find the power-series expansion for the function. (This method is due to Newton.)

3.8 ALTERNATIVE PROCEDURE

The power-series solution of an equation such as (7.1) can be constructed in another way. Instead of assuming a form (7.2) and substituting into the differential equations to obtain relations to determine the unknown coefficients a_0, a_1, etc., we can determine these from the formula

$$a_n = \frac{u^{(n)}(0)}{n!}. \tag{8.1}$$

Referring to (7.1), we are given $a_0 = u(0) = 1$, $a_1 = u'(0) = 0$ to begin with. To compute the higher-order derivatives, we use the differential equation itself. Thus,

$$\begin{aligned} u'' &= xu, \\ u''' &= xu' + u, \\ u^{iv} &= xu'' + 2u', \\ u^{v} &= xu''' + 3u''. \end{aligned} \tag{8.2}$$

In general,

$$u^{(n)} = xu^{(n-2)} + (n-2)u^{(n-3)}, \qquad n = 2, 3, \ldots. \tag{8.3}$$

Therefore,

$$u^{(n)}(0) = (n-2)u^{(n-3)}(0). \tag{8.4}$$

From this, we readily obtain

$$\begin{aligned} u''(0) &= u^{(5)}(0) = \cdots = 0, \\ u^{(4)}(0) &= u^{(7)}(0) = \cdots = 0, \\ u'''(0) &= u(0) = 1, \qquad u^{(6)}(0) = 4u'''(0) = 4, \ldots. \end{aligned} \tag{8.5}$$

Therefore

$$a_3 = \frac{1}{3!}, \qquad a_6 = \frac{4}{6!}, \qquad a_9 = \frac{7 \cdot 4}{9!}, \qquad \ldots, \tag{8.6}$$

which yields (7.8) as before. The same procedure can be used to derive (7.9).

EXERCISES

1. Use the method of this section to find the principal solutions of each of the following equations.

   a) $u'' - u = 0$ b) $u'' - xu' - u = 0$

   c) $u'' + 4x^2u = 0$ d) $u'' - 2xu' + 2u = 0$.

2. The equation

$$(1 - x^2)u'' - 2xu' + n(n+1)u = 0,$$

   called the *Legendre differential equation*, occurs in mathematical physics.

   a) Determine the principal solutions.

   b) Show that if n is a nonnegative integer, one of the principal solutions is a polynomial. Find this polynomial for $n = 0, 1, 2,$ and 3.

3.9 CONVERGENCE (1)

It is easy, in the case of the series obtained in Section 3.7, to apply the standard ratio test and show that the series converge for all x. What is the general situation? Pursuant to our general policy, we shall avoid any detailed analysis which takes us too far afield and merely state the essential results. Proofs will be found in more advanced texts referred to at the end of the chapter.

A most important result is the following theorem.

Theorem. *Consider the linear differential equation*

$$u'' + p_1(x)u' + p_2(x)u = 0, \tag{9.1}$$

where $p_1(x)$ and $p_2(x)$ are power series convergent in $|x| < b$:

$$p_1(x) = b_0 + b_1x + \cdots, \qquad p_2(x) = c_0 + c_1x + \cdots \tag{9.2}$$

Then every solution of Eq. (9.1) *can be represented as a power-series convergent in $|x| < b$.*

Corollary. *If $p_1(x)$ and $p_2(x)$ are polynomials, then every solution of Eq.* (9.1) *can be represented as a power series which converges everywhere.*

The above corollary enables us to look at a linear differential equation such as

$$u'' + 2xu' + (1 + x^3)u = 0, \tag{9.3}$$

and immediately conclude that we can apply the power-series technique with complete confidence in the validity of our results for all x.

Let us point out that the corresponding statement is valid for linear differential equations of any order and for linear systems. We shall discuss this topic again in Chapter 5.

EXERCISE

1. What can be said, on the basis of the above theorem, about the radius of convergence of the series solutions of the following equations?

   a) $u'' - x^2u = 0$

   b) The equation in Exercise 6 of Section 3.7

   c) The Legendre differential equation

   d) $u'' + (\sin x)u = 0$

   e) $e^x u'' + x^2u = 0$

   f) $(1 + x^3)u'' + u = 0$

3.10 EXPANSIONS ABOUT $x = a$

In all of the examples above, the initial conditions have been given at $x = 0$; that is, $u(0)$ and $u'(0)$ have been given, and the solution has been sought as a power series in x of the form

$$u = a_0 + a_1 x + a_2 x^2 + \cdots. \tag{10.1}$$

In some instances, the initial conditions may be at some other value of x, say $x = a$. That is, the values $u(a)$ and $u'(a)$ may be given. In such a case it is natural to seek a solution in the form of a series of powers of $x - a$, a Taylor expansion about the point $x = a$,

$$u = a_0 + a_1(x - a) + a_2(x - a)^2 + \cdots. \tag{10.2}$$

Let us illustrate this for the problem

$$u'' - xu = 0, \qquad u(1) = 1, \qquad u'(1) = 0. \tag{10.3}$$

Assume that there is a solution

$$u = a_0 + a_1(x - 1) + a_2(x - 1)^2 + \cdots + a_n(x - 1)^n + \cdots, \tag{10.4}$$

convergent in some interval $|x - 1| < b$. Then

$$\begin{aligned} u' &= a_1 + 2a_2(x - 1) + \cdots + na_n(x - 1)^{n-1} + \cdots, \\ u'' &= 2a_2 + 6a_3(x - 1) + \cdots + n(n - 1)a_n(x - 1)^{n-2} + \cdots. \end{aligned} \tag{10.5}$$

In order to express $u'' - xu$ in powers of $x - 1$, we write

$$\begin{aligned} u'' - xu &= u'' - (x - 1)u - u \\ &= (2a_2 - a_0) + (6a_3 - a_1 - a_0)(x - 1) + \cdots \\ &\quad + [n(n - 1)a_n - a_{n-2} - a_{n-3}](x - 1)^{n-2} + \cdots. \end{aligned} \tag{10.6}$$

In order that this should be zero for $|x - 1| < b$ it is necessary that

$$\begin{aligned} 2a_2 - a_0 = 0, 6a_3 - a_1 - a_0 = 0, \\ n(n - 1)a_n - a_{n-2} - a_{n-3} = 0. \end{aligned} \tag{10.7}$$

From (10.3) it follows that $a_0 = 1$, $a_1 = 0$, and then from (10.7) we have

$$\begin{aligned} a_2 = \frac{1}{2}, \qquad a_3 = \frac{1}{6} = \frac{1}{2 \cdot 3}, \qquad a_4 = \frac{a_2 + a_1}{4 \cdot 3} = \frac{1}{4 \cdot 3 \cdot 2}, \\ a_5 = \frac{a_3 + a_2}{5 \cdot 4} = \frac{4}{5!}. \end{aligned} \tag{10.8}$$

Thus we obtain

(10.9)

$$u = 1 + \frac{1}{2!}(x-1)^2 + \frac{1}{3!}(x-1)^3 + \frac{1}{4!}(x-1)^4 + \frac{4}{5!}(x-1)^5 + \cdots.$$

In this example, it is not readily apparent what the general coefficient in the series is. This will often be the case when the *recurrence relation* expresses a_n in terms of more than one previous coefficient, as for example in (10.7).

It is again true that the coefficients in the series (10.2) can be computed directly from the differential equation by successive differentiations, as in Section 3.8. This time the basic formula is

$$a_n = \frac{u^{(n)}(a)}{n!}. \tag{10.10}$$

For the problem (10.3), for example, the derivatives are given by (8.2) and (8.3), from which one gets

$$\begin{aligned} u''(1) &= u(1) = 1, \\ u'''(1) &= u'(1) + u(1) = 1, \\ u^{iv}(1) &= u''(1) + 2u'(1) = 1, \\ u^{v}(1) &= u'''(1) + 3u''(1) = 4, \end{aligned} \tag{10.11}$$

and so on.

Still another procedure is possible, namely to translate the x-coordinate so that $x = a$ becomes the origin of a new coordinate system. For example, for the problem (10.3) the substitution $x - 1 = z$ leads to the problem

$$\frac{d^2u}{dz^2} - (z+1)u = 0, \qquad u = 1 \text{ at } z = 0, \qquad u' = 0 \text{ at } z = 0. \tag{10.12}$$

Now a solution can be obtained in powers of z,

$$u = a_0 + a_1 z + a_2 z^2 + \cdots, \tag{10.13}$$

using the previous techniques. Replacement of z by $x - 1$ in this series yields the desired series.

EXERCISES

1. Find a series solution for

$$u'' - xu = 0, \qquad u(1) = 0, \qquad u'(1) = 1.$$

2. Find the first few terms in two linearly independent power series solutions in powers of $x - a$ for each of the following equations.

   a) $u'' - u = 0, \qquad a = \pi/4,$

   b) $u'' - xu' - u = 0, \qquad a = 1,$

   c) $u'' + (\sin x)u = 0, \qquad a = \pi/2,$

   d) $u'' + (\ln x)u = 0, \qquad a = 1.$

3. Reformulate the convergence theorem of Section 3.9 so that it applies to expansions in powers of $x - a$. On the basis of the result, what can be said about the radius of convergence of the series solutions in Exercise 2?

3.11 NONLINEAR EQUATIONS

The same techniques permit us to obtain the solution of nonlinear equations. Consider, for example, the Riccati equation

$$u' = x + u^2, \qquad u(0) = 1. \tag{11.1}$$

As before, we set

$$u = 1 + a_1x + a_2x^2 + \cdots, \tag{11.2}$$

and substitute in Eq. (11.1). We may just as well use the condition $u(0) = 1$ immediately to deduce that $a_0 = 1$. The result is

(11.3)
$$\begin{aligned} a_1 + 2a_2x + 3a_3x^2 + \cdots &= x + (1 + a_1x + a_2x^2 + \cdots)^2 \\ &= x + [1 + 2a_1x + (a_1^2 + 2a_2)x^2 + \cdots]. \end{aligned}$$

Equating coefficients, we obtain the following relations:

$$a_1 = 1, \qquad 2a_2 = 1 + 2a_1, \qquad 3a_3 = a_1^2 + 2a_2, \tag{11.4}$$

and so on.

The point is that, just as in the linear case, we obtain recurrence relations which permit us to determine the coefficients in the power series. These relations are now nonlinear and not at all as simple as before. For our purposes, however, the important point is that they can be determined one at a time in terms of the previous coefficients. A result concerning convergence is given in Section 3.15.

EXERCISES

1. Obtain a power-series expansion for $\tan x$ using the differential equation $u' = 1 + u^2$, $u(0) = 0$.

2. Obtain a power series for $(e^x - e^{-x})/(e^x + e^{-x})$ using a corresponding differential equation.

3. Obtain a power-series expansion for $u(x) = \log \cos x$.

4. Let $u'' - xu = 0$, $u(0) = 1$, $u'(0) = 0$. Obtain the power series expansion for $\log u$. [*Hint:* $v = u'/u$ is the solution of the Riccati equation $v' + v^2 - x = 0$, $v(0) = 0$.]

3.12 NUMERICAL ASPECTS

Power series are extremely convenient as a means of obtaining numerical values of functions. Suppose, for example, we wish to evaluate $\sin x$ for various values of x such as $\pi/6$ and $\pi/3$. We start with the power series

$$\sin x = x - \frac{x^3}{6} + \frac{x^5}{120} - \cdots \tag{12.1}$$

Using the approximate value $\pi \cong 3.1416$, we have

$$x = \pi/6 = 0.5236, \qquad x^3 = 0.14354, \qquad x^5 = 0.03935. \tag{12.2}$$

We see that the contribution of the term $x^5/120$ is less than four parts in 10^4. Hence we can ignore it if we wish the value of $\sin(\pi/6)$ to, say, three significant figures.

How do we know that the sum of all the remaining terms in the power series

$$\frac{x^5}{120} - \frac{x^7}{5050} + \cdots \tag{12.3}$$

will not add up to a contribution that is significant? This is a valid question and one that must always be kept in mind. In this case, and in many cases, various estimation techniques can be used to bound the error. In many other cases, we proceed on faith.

An easily applied result is the following: If $a_0 > a_1 > a_2 > \cdots$ and a_n approaches zero as $n \to \infty$, then the series

$$S = a_0 - a_1 + a_2 - a_3 + \cdots \tag{12.4}$$

converges; and, for any n,

$$(12.5)\qquad a_0 - a_1 + a_2 - a_3 + \cdots + a_{2n} > S > a_0 - a_1 + a_2 - a_3 + \cdots + a_{2n} - a_{2n+1}.$$

From this inequality we see that we can obtain an estimate of the truncation error involved in using $S \cong S_n = a_0 - a_1 + \cdots \pm a_n$ in terms of the first term neglected.

Proceeding as above, we have

$$(12.6)\qquad \sin\frac{\pi}{6} \cong \frac{\pi}{6} - \frac{(\pi/6)^3}{6} = 0.4997.$$

This value compares favorably with the exact value of $\sin(\pi/6)$, namely $\frac{1}{2} = 0.5000$.

If we wished to compute $\sin(\pi/3)$ in this fashion, we would need to use the term involving $x^5/120$. In practice, we would use a double-angle formula,

$$(12.7)\qquad \sin 2x = 2\sin x\cos x,$$

to reduce the value of x, or any of a number of other methods which involve considerably less arithmetic. In the days before desk computers and digital computers, questions as to the most adroit way of calculating functional values of even elementary functions occupied the greatest mathematicians. Calculation is as important today, but our attention has shifted to more complicated functions and more sophisticated algorithms.

EXERCISES

1. Obtain the value of $\cos(\pi/6)$ from that of $\sin(\pi/6)$, using the relation

$$\sin^2 x + \cos^2 x = 1.$$

2. Obtain the value of $\cos(\pi/6)$ using the power series

$$\cos x = 1 - \frac{x^2}{2} + \frac{x^4}{24} - \cdots,$$

   and compare the two values thus obtained with the exact value $\sqrt{3}/2$.

3. Evaluate $\sin(\pi/18)$ using the power series for $\sin x$, and then evaluate $\sin(\pi/6)$ using the expression for $\sin 3x$ in terms of $\sin x$. Compare the value thus obtained with the previous two values.

4. Consider the power series

$$\log(1+x) = x - \frac{x^2}{2} + \frac{x^3}{3} - \cdots$$

The series converges for $x = 1$. How many terms would one need to compute $\log 2$ to an accuracy of one in 10^2?

5. Consider the series

$$\log \frac{1+x}{1-x} = 2x + \frac{2x^3}{3} + \frac{2x^5}{5} + \cdots$$

If $x = \frac{1}{3}$, then $(1+x)/(1-x) = 2$. Hence

$$\log 2 = 2(\tfrac{1}{3}) + \tfrac{2}{3}(\tfrac{1}{3})^3 + \tfrac{2}{5}(\tfrac{1}{3})^5 + \cdots$$

How many terms are required to calculate $\log 2$ to an accuracy of one in 10^2?

6. Calculate the values of $u(x)$, the solution of

$$u'' - x^2 u = 0, \qquad u(0) = 1, \qquad u'(0) = 0,$$

at $x = 0.1, 0.5, 1$ to accuracies of one in 10^3.

7. For what value of x is $u(x)$ equal to 1.1? Obtain this value to an accuracy of one in 10^3.

3.13 THE SWINGING PENDULUM(1)

Under reasonable assumptions, we can describe the motion of a pendulum initially at rest and started with a velocity v by means of the equation

$$my'' + g \sin y = 0, \qquad y(0) = 0, \qquad y'(0) = v. \tag{13.1}$$

(See Fig. 3.1.) There is a simple technique for obtaining the solution in terms of elliptic functions. For pedagogical purposes, we shall pretend that we do not know this. In any case, it would require a great deal of effort to introduce

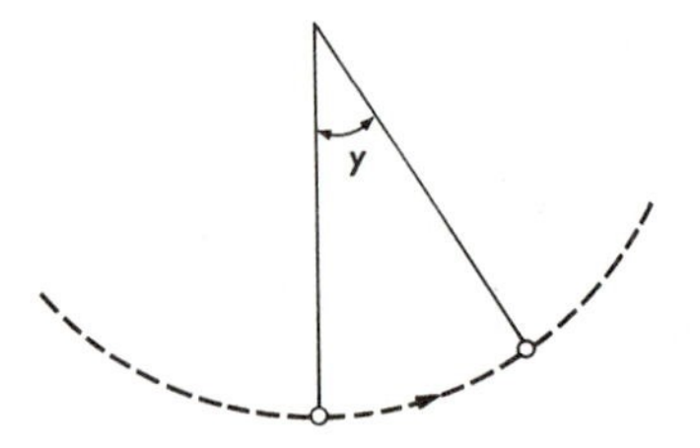

Figure 3.1

and study these functions. At the moment, we want a simple way of obtaining some numerical results concerning the behavior of the pendulum. To illustrate the basic ideas, let us suppose that we have changed the angle and time coordinates so that the equation reads

$$y'' + \sin y = 0, \qquad y(0) = 0, \qquad y'(0) = \tfrac{1}{2}. \tag{13.2}$$

We would like to determine the maximum amplitude and the period of oscillation. We have not, of course, established that this equation gives rise to a periodic oscillation. Our objective will be to discover the nature of the oscillation, if such an oscillation takes place.

We have already seen that $\sin y \cong y$ for $|y| \leq \frac{1}{2}$. Let us suppose that the initial velocity will be decreased continually by the gravitational effect until the velocity is zero at maximum amplitude. In physical terms, kinetic energy is being transformed into potential energy. In any case, we consider the approximate equation

$$z'' + z = 0, \qquad z(0) = 0, \qquad z'(0) = \tfrac{1}{2}, \tag{13.3}$$

and obtain the explicit solution $z = (\sin x)/2$. The maximum amplitude is $\frac{1}{2}$, attained at $x = \pi/2$, and the period of oscillation is 2π. What happens in the more realistic description of Eq. (13.2)?

Let us begin by seeing how well we could answer these questions for Eq. (13.3), using a power-series representation for z in terms of x. We have

$$z = \frac{\sin x}{2} = \frac{1}{2}\left(x - \frac{x^3}{6} + \frac{x^5}{120} - \cdots\right). \tag{13.4}$$

Let us use the first three terms:

$$z \cong \frac{1}{2}\left(x - \frac{x^3}{6} + \frac{x^5}{120}\right). \tag{13.5}$$

The maximum amplitude is attained at $z' = 0$:

$$0 = 1 - \frac{x^2}{2} + \frac{x^4}{24}. \tag{13.6}$$

As a quadratic equation in x^2, Eq. (13.6) has the two roots

$$x^2 = 6 - 2\sqrt{3}, \qquad x^2 = 6 + 2\sqrt{3}. \tag{13.7}$$

Using the value $\sqrt{3} \cong 1.732$, we have

$$x^2 \cong 2.536, \qquad x \cong 1.592. \tag{13.8}$$

This value compares quite favorably with $\pi/2 \cong 1.571$. The other root in (13.7) yields a value which compares well with $\pi \cong 3.142$, considering the crudeness of the approximation we are employing.

EXERCISE

1. Use the first four terms of the expansion to find an improved estimate for $\pi/2$. [*Hint:* The root of the resulting cubic equation in x^2 is close to $\pi/2$, which is to say it is close to the approximate value already found. Use the Newton-Raphson approximation method to find an improved value.]

3.14 THE SWINGING PENDULUM(2)

Let us now turn to Eq. (13.2). We write

$$y = \tfrac{1}{2}(x + a_3x^3 + a_5x^5 + \cdots). \tag{14.1}$$

Then

$$y'' = \tfrac{1}{2}(6a_3x + 20a_5x^3 + \cdots) \tag{14.2a}$$

and

$$\begin{aligned} \sin y &= y - \frac{y^3}{6} + \cdots \\ &= \tfrac{1}{2}(x + a_3x^3 + a_5x^5) - \frac{(x + a_3x^3 + a_5x^5)^3}{48} + \cdots \end{aligned} \tag{14.2b}$$

In order to determine a_3 and a_5, we need only keep the terms in x and x^3 on the right-hand side of (14.2b). Upon grouping, we see that

$$\sin y = \frac{1}{2}x + x^3\left(\frac{a_3}{2} - \frac{1}{48}\right) + \cdots \tag{14.3}$$

Hence, upon equating coefficients of x and x^3, we see that the equation

$$y'' + \sin y = 0 \tag{14.4}$$

yields the relations

$$3a_3 + \tfrac{1}{2} = 0, \qquad 10a_5 + \frac{a_3}{2} - \frac{1}{48} = 0. \tag{14.5}$$

Thus

(14.6) $$a_3 = -\tfrac{1}{6}, \qquad a_5 = \tfrac{1}{96}.$$

The power-series solution of Eq. (13.2) has the form

(14.7) $$y = \frac{1}{2}\left(x - \frac{x^3}{6} + \frac{x^5}{96}\cdots\right).$$

Comparing (14.7) with the series for $\sin x$, we see that consideration of the nonlinear term $-y^3/6$ produces a very small change for small x.

EXERCISES

1. Using the method just outlined, obtain the coefficients of x^7 and x^9 in the power-series expansion of the solution of Eq. (13.2).
2. Use the first three terms, and then the first five terms to determine the period and the maximum amplitude.
3. For what time will y assume the value $\frac{1}{4}$?

3.15 CONVERGENCE(2)

In the foregoing sections we illustrated the use of power-series expansion to obtain numerical results. Various expansions were manipulated formally with no attention to convergence. What can we say about convergence?

The most important statement is that the foregoing techniques are valid whenever the solution possesses a power-series expansion and we stay within the interval of convergence. How do we know, however, that such an expansion exists, and how do we determine what the interval of convergence is?

It must be stated clearly that there is no systematic way of answering these questions. Sometimes we are guided by general theorems, sometimes by particular features of the equation under consideration, sometimes by physical intuition, and, unfortunately, quite often by trial and error.

A very useful result is the following:

Theorem. *Consider the equation*

(15.1) $$u' = p(x, u), \qquad u(0) = c,$$

where $p(x, u)$ is either a polynomial in x and u or, more generally, a power series in u, with coefficients which are polynomials in x, convergent for $|u| < b$, *where $b > |c|$. Then in some interval $|x| < a$ this equation has a unique solution which is expressible as a power series:*

$$u = c + a_1x + a_2x^2 + \cdots \tag{15.2}$$

The coefficients can be obtained recurrently by substitution in (15.1) *and equating of coefficients.*

Corresponding results hold for second-order equations of the form

$$u'' = p(x, u, u'), \qquad u(0) = c_1, \qquad u'(0) = c_2. \tag{15.3}$$

There is no general method known for determining the value of the radius of convergence a. The foregoing theorem enables us to ascertain at a glance that equations such as

$$u' = \frac{x + u^2}{1 + u^2}, \qquad u(0) = \frac{1}{2}, \tag{15.4}$$

$$u' = \frac{u^2 + 1}{1 - x}, \qquad u(0) = 1,$$

possess power-series solutions for sufficiently small $|x|$. On the other hand, equations such as

$$u' = \frac{u^2}{x + u}, \qquad u(0) = 1, \tag{15.5}$$

$$x^2u' = 1 + u^2, \qquad u(0) = 2,$$

require more sophisticated analysis and, indeed, may not even possess power-series solutions. It is fair to warn the reader that the area is one of great intricacy and that differential equations that arise in applications frequently require quite advanced techniques plus analytic ingenuity. Our aim here has been to provide the reader with some useful preliminary tools and to motivate his further study.

Analogous results for higher-order equations are discussed in Chapter 5.

EXERCISES

1. Obtain the first four terms of the power-series expansions of the solutions of the differential equations (15.4).

2. Consider the Riccati equation

$$u' = u^2 + 1, \qquad u(0) = 1.$$

Obtain the first four terms of the power-series expansion and show that the radius of convergence is $\pi/2$. [*Hint:* $u = \tan x = \sin x/\cos x$.]

3. Obtain the power-series solution of

$$u' = -u^2 + x, \qquad u(0) = 1,$$

by using the fact that $u = v'/v$, where v satisfies the equation

$$v'' - xv = 0.$$

4. Use a modification of the idea used in Exercise 3 to obtain the power-series solution of

$$u' = u^2 + xu + 1, \qquad u(0) = 1.$$

Obtain an approximate value for the radius of convergence.

5. In general, what can one say about the radius of convergence of the power series for the solution of

$$u' + u^2 = p(x)u + q(x), \qquad u(0) = 1,$$

where p and q are polynomials in x?

3.16 SLIGHTLY MORE GENERAL LINEAR DIFFERENTIAL EQUATIONS

Despite the general difficulties about which the reader was warned in the last section, we may reasonably expect to make some further progress by considering linear differential equations. Consider, for example, the equation

$$x^2u'' - (1 + x)u = 0. \tag{16.1}$$

Because of the coefficient x^2, this equation does not fit into any of our preceding categories of relatively tame specimens of either linear or nonlinear equations. In fact, on comparing it with Eq. (9.1), we see that $p_1(x) = 0$, $p_2(x) = (1 + x)/x^2$. Since the function $p_2(x)$ does not have a convergent expansion in powers of x, the convergence theorem cannot be invoked to assure us that a solution which exists near $x = 0$ can be represented as a power series in x. In such a case, the point $x = 0$ is called a *singular point*. Let us nonetheless try to determine a solution of the form

$$u = a_0 + a_1x + a_2x^2 + \cdots \tag{16.2}$$

Differentiating (16.2), we have

$$u' = a_1 + 2a_2x + 3a_3x^2 + \cdots, \quad u'' = 2a_2 + 6a_3x + 12a_4x^2 + \cdots, \tag{16.3}$$

and thus

$$x^2u'' - (1 + x)u = 2a_2x^2 + 6a_3x^3 + 12a_4x^4 + \cdots - a_0 - (a_1 + a_0)x - (a_2 + a_1)x^2 - \cdots \tag{16.4}$$

Equating coefficients to satisfy (16.1) yields

$$\begin{aligned} -a_0 &= 0, \\ -(a_1 + a_0) &= 0, \\ 2a_2 - (a_2 + a_1) &= 0, \\ &\vdots \\ n(n-1)a_n - (a_n + a_{n-1}) &= 0. \end{aligned} \tag{16.5}$$

We see, recursively, that

$$a_0 = a_1 = a_2 = \cdots = 0, \tag{16.6}$$

a very unsatisfactory state of affairs!

3.17 DISCUSSION

It is clear that a direct attempt to find a power-series solution has failed. What do we do next? In order to get a hint as to how to proceed, let us examine the particular equation

$$x^2u'' - u = 0. \tag{17.1}$$

Referring to Section 2.33 of Chapter 2, we know that there are two solutions of Eq. (17.1) of the form

$$u = x^\rho, \tag{17.2}$$

where ρ can be either of the roots of the quadratic equation

$$\rho(\rho - 1) - 1 = 0. \tag{17.3}$$

Guided by this particular case, let us try a solution of Eq. (17.1) possessing the form

$$u = x^\rho(1 + a_1x + a_2x^2 + \cdots), \tag{17.4}$$

a modified power-series expansion.

Before embarking on this task, let us once again emphasize that the subject of power-series solutions of linear differential equations is both difficult and complex, and that the techniques must be used with great care to avoid egregious errors. We shall therefore content ourselves with an indication of how to obtain the simplest, more important, and most frequently used results and refer the reader to more advanced works for detailed discussions and extensions.

3.18 MODIFIED POWER-SERIES SOLUTION

It seems reasonable, in view of what we said in the last section, to look for classes of linear differential equations which permit solutions of the form

$$u = x^\rho[a_0 + a_1x + a_2x^2 + \cdots], \tag{18.1}$$

where ρ is a parameter to be determined, along with the coefficients a_i.

To illustrate the applicability of these more general power series, consider the equation

$$x^2u'' - (1 + x)u = 0, \tag{18.2}$$

already encountered in Section 3.16. We have

$$\begin{aligned} u &= a_0x^\rho + a_1x^{\rho+1} + a_2x^{\rho+2} + \cdots, \\ u' &= \rho a_0x^{\rho-1} + (\rho + 1)a_1x^\rho + (\rho + 2)a_2x^{\rho+1} + \cdots, \\ u'' &= \rho(\rho - 1)a_0x^{\rho-2} + (\rho + 1)\rho a_1x^{\rho-1} \\ &\qquad + (\rho + 2)(\rho + 1)a_2x^\rho + \cdots, \\ (1 + x)u &= a_0x^\rho + (a_1 + a_0)x^{\rho+1} + (a_2 + a_1)x^{\rho+2} + \cdots \end{aligned} \tag{18.3}$$

Hence, collecting terms, we have

(18.4)

$$\begin{aligned} x^2u'' - (1 + x)u &= [\rho(\rho - 1) - 1]a_0x^\rho \\ &\quad + \{[(\rho + 1)\rho - 1]a_1 - a_0\}x^{\rho+1} + \cdots \\ &\quad + \{[(\rho + n)(\rho + n - 1) - 1]a_n - a_{n-1}\}x^{\rho+n} + \cdots \end{aligned}$$

The right-hand side of (18.4) will be identically zero if all the coefficients are zero. The conditions are

$$\begin{aligned} [\rho(\rho - 1) - 1]a_0 &= 0, \\ [(\rho + 1)\rho - 1]a_1 - a_0 &= 0, \\ &\vdots \\ [(\rho + n)(\rho + n - 1) - 1]a_n - a_{n-1} &= 0. \end{aligned} \tag{18.5}$$

We may just as well suppose that $a_0 \neq 0$; for if a_k is the first nonzero coefficient, then we have merely shifted the value of ρ by k. Hence, if $a_0 \neq 0$, we must have

$$\rho(\rho - 1) - 1 = 0. \tag{18.6}$$

This is an important equation, the *indicial equation*. Let the two roots be denoted by ρ_1 and ρ_2. Then

$$\rho_1 = \frac{1 + \sqrt{5}}{2}, \qquad \rho_2 = \frac{1 - \sqrt{5}}{2}. \tag{18.7}$$

If $a_0 \neq 0$, we may as well set $a_0 = 1$, since the equation is linear. The second equation in (18.5) yields

$$a_1 = \frac{1}{(\rho + 1)\rho - 1}, \tag{18.8}$$

provided that the denominator is nonzero. Since

$$(\rho + 1)\rho - 1 = \rho^2 + \rho - 1 = (\rho + 1) + \rho - 1 = 2\rho \tag{18.9}$$

(using the fact that $\rho^2 = \rho + 1$), we see that the denominator is nonzero. Similarly, we see that

$$\begin{aligned} a_2 &= \frac{a_1}{(\rho + 2)(\rho + 1) - 1} \\ &= \frac{1}{[(\rho + 2)(\rho + 1) - 1][(\rho + 1)\rho - 1]} \end{aligned} \tag{18.10}$$

where none of the terms in the denominator is zero, and we can determine the coefficients one after another as before.

We see then that there exist two solutions of Eq. (18.2), each having the form

$$\begin{aligned} u = x^\rho \Bigg[1 &+ \frac{x}{[(\rho + 1)\rho - 1]} \\ &+ \frac{x^2}{[(\rho + 2)(\rho + 1) - 1][(\rho + 1)\rho - 1]} + \cdots \Bigg], \end{aligned} \tag{18.11}$$

where ρ is one of the values in (18.7).

These two solutions are distinct, and we should expect that the general solution is a linear combination of these particular solutions. This is indeed the case, but we are not prepared to prove it at this stage.

EXERCISES

1. For each equation, find the indicial equation and its roots and find the modified series solution corresponding to each root.
   a) $2xu'' + u' + xu = 0$
   b) $3x^2u'' + 2xu' + x^2u = 0$
   c) $2x^2u'' - (5 \sin x)u' + 3u = 0$.
2. Consider the equation

$$(*) \qquad x^2y'' + xy' + (x^2 - n^2)y = 0$$

   (Bessel's equation). Show that the indicial equation has the roots $\rho = \pm n$.
3. For the case $n = 0$, show that Eq. (*) has a solution of the form

$$y = 1 - \frac{x^2}{2^2(1!)^2} + \frac{x^4}{2^4(2!)^2} - \frac{x^6}{2^6(3!)^2} + \cdots$$

   This series is called the Bessel function of order zero and is written $J_0(x)$. Show directly that the series converges for all real x and thus that it is rigorously a solution of the foregoing differential equation.
4. For the case $n = 1$, show that Eq. (*) has a solution of the form

$$y = \frac{x}{2} - \frac{x^3}{2^3 1!2!} + \frac{x^5}{2^5 2!3!} - \cdots$$

   Establish results corresponding to those of Exercise 3. Calculate the numerical values of $J_0(0.1)$, $J_0(1)$ to various degrees of accuracy.
5. What does the solution corresponding to the indicial root $\rho = -1$ look like?
6. Obtain the modified power-series expansion for the case $n = \frac{1}{2}$ and compare it with the function $(\sin x)/\sqrt{x}$.
7. Consider the differential equation,

$$x(1 - x)u'' + [c - (a + b + 1)x]u' - abu = 0,$$

the famous hypergeometric equation of Gauss. Show that there is a power-series solution of the form

$$u = 1 + \frac{ab}{c}x + \frac{a(a+1)b(b+1)}{2!c(c+1)}x^2 + \frac{a(a+1)(a+2)b(b+1)(b+2)}{3!c(c+1)(c+2)}x^3 + \cdots$$

8. Consider the differential equation $(1 - x^2)u'' - 2xu' + n(n + 1)u = 0$. Show that if n is an integer, there is a polynomial solution of degree n, the Legendre polynomial, and find the expressions for the coefficients.

9. Show that

$$xy'' - (x + n)y' + ny = 0$$

has the particular solutions e^x and $1 + x/1! + x^2/2! + \cdots + x^n/n!$. Hence, show that the solution is $x^{n+1}/(n + 1)! + x^{n+2}/(n + 2)! + \cdots$.

3.19 STILL MORE GENERAL LINEAR DIFFERENTIAL EQUATIONS

What happens if the coefficient of the second derivative is a polynomial of degree higher than two? The answer is that the situation becomes even more complicated rather rapidly. There are systematic techniques available to handle general linear differential equations of the form

$$p_1(x)y'' + p_2(x)y' + p_3(x)y = 0, \tag{19.1}$$

but one could easily spend an entire semester on the detailed investigation that is required.

To indicate the kind of difficulty that is encountered, consider the equation

$$x^3u' - 2u = 0, \qquad u(0) = 1. \tag{19.2}$$

It has the solution

$$u = e^{-1/x^2}. \tag{19.3}$$

The function e^{-1/x^2} is a well-known example of a function possessing all derivatives at $x = 0$ and yet possessing no power-series expansion.

It is important to emphasize that the question of effective numerical solution of equations such as (19.2), and, more generally, Eq. (19.1), is one that requires a high level of analytic effort. They are prime examples of equations which cannot be blithely delegated to a digital computer for routine calculation.

Although we shall emphasize in the next chapter how easy it is to treat certain classes of differential equations in a simple fashion, we wish constantly to point out the need for mathematical training which enables the scientist to recognize these classes of equations. The overall problem is that of using the appropriate method at the appropriate time.

3.20 PERTURBATION TECHNIQUES

Consider the equation

$$u'' + (1 + \epsilon x)u = 0, \qquad u(0) = 1, \qquad u'(0) = 0, \tag{20.1}$$

where ϵ is a constant. Using the methods of the previous sections, we readily obtain the power-series solution

$$u = 1 - \frac{x^2}{2} - \frac{\epsilon x^3}{6} + \cdots \tag{20.2}$$

Suppose that ϵ is 0.01 and that we want to evaluate the solution at $x = 5$. Despite the fact that the series converges for all values of x, as we indicated in Section 3.3, we see that the calculation for $x = 5$ is a considerable task. The power-series expansion must be calculated for quite a number of terms before a numerical value with any accuracy can be obtained.

It is clear that an alternative approach is desirable. Observe that the term ϵx is small in absolute value for $0 \leq x \leq 5$. In a reasonable universe, therefore, it would be true that the solution of Eq. (20.1) would differ from that of

$$v'' + v = 0, \qquad v(0) = 1, \qquad v'(0) = 0, \tag{20.3}$$

by a small amount. Let us act upon that hypothesis. Subsequently, we will indicate certain dangers in accepting this premise unreservedly.

Instead of using a power series in x, which yields a solution useful for numerical purposes only in the neighborhood of $x = 0$, let us expand the solution as a power series in the parameter ϵ. We thus obtain a simple equation with an explicit solution when $\epsilon = 0$. Write

$$u = u_0 + \epsilon u_1 + \epsilon^2 u_2 + \cdots, \tag{20.4}$$

where the quantities $u_0, u_1, u_2, \ldots,$ are functions of x to be determined. The series (20.4) is called a *perturbation series*. Substituting (20.4) in (20.1), we have

$$(u_0'' + \epsilon u_1'' + \epsilon^2 u_2'' + \cdots) + (1 + \epsilon x)(u_0 + \epsilon u_1 + \epsilon^2 u_2 + \cdots) = 0. \tag{20.5}$$

Collecting coefficients of powers of ϵ, we obtain

$$(u_0'' + u_0) + \epsilon(u_1'' + u_1 + xu_0) + \epsilon^2(u_2'' + u_2 + xu_1) + \cdots = 0. \tag{20.6}$$

A simple way to satisfy this equation is to suppose that $u_0, u_1, u_2, \ldots,$ are independent of ϵ and to equate the coefficients of the various powers of ϵ to zero. We thus obtain the sequence of equations

$$(20.7)\qquad \begin{aligned} u_0'' + u_0 &= 0,\\ u_1'' + u_1 + xu_0 &= 0,\\ u_2'' + u_2 + xu_1 &= 0, \quad \text{etc.} \end{aligned}$$

How do we determine the initial conditions? Setting $x = 0$ in the expression (20.4) before and after differentiation, we obtain the two relations

$$(20.8)\qquad \begin{aligned} u_0(0) + \epsilon u_1(0) + \epsilon^2 u_2(0) + \cdots &= 1,\\ u_0'(0) + \epsilon u_1'(0) + \epsilon^2 u_2'(0) + \cdots &= 0. \end{aligned}$$

Equating the coefficients of like powers of ϵ on both sides, we obtain the initial conditions

$$(20.9)\qquad \begin{aligned} u_0(0) &= 1, & u_0'(0) &= 0,\\ u_1(0) &= 0, & u_1'(0) &= 0,\\ u_2(0) &= 0, & u_2'(0) &= 0. \end{aligned}$$

Thus the equation for u_0 is

$$(20.10)\qquad u_0'' + u_0 = 0, \qquad u_0(0) = 1, \qquad u_0'(0) = 0,$$

while those for $u_1, u_2, \ldots,$ are all of the same form:

$$(20.11)\qquad \begin{aligned} u_1'' + u_1 + xu_0 &= 0, & u_1(0) = u_1'(0) &= 0,\\ u_2'' + u_2 + xu_1 &= 0, & u_2(0) = u_2'(0) &= 0, \quad \text{etc.} \end{aligned}$$

3.21 ANALYTIC EXPRESSIONS

The solution to Eq. (20.10) is readily seen to be $u_0 = \cos x$. Using this expression in Eqs. (20.11), we find that the equation for u_1 is

$$(21.1)\qquad u_1'' + u_1 + x\cos x = 0, \qquad u_1(0) = 0, \qquad u_1'(0) = 0.$$

From (21.1) and using the techniques of Chapter 2, we readily obtain

$$(21.2)\qquad u_1 = \frac{(1 - x^2)\sin x - x\cos x}{4}.$$

Substituting this expression in Eqs. (20.11), we obtain

$$(21.3)\qquad u_2'' + u_2 + x\left[\frac{(1 - x^2)\sin x - x\cos x}{4}\right] = 0,$$

$$u_2(0) = 0, \qquad u_2'(0) = 0,$$

thus determining u_2. By similar procedures we can find the values of the other u_i.

Thus an approximate expression for u is

$$u \cong \cos x + \epsilon \left[\frac{(1 - x^2) \sin x - x \cos x}{4} \right]. \tag{21.4}$$

The expression (21.4) is accurate so long as ϵx^2 is small compared to 1. Therefore, if $\epsilon = 0.01$, (21.4) would give a good approximation for $0 \leq x \leq 4$.

EXERCISES

1. Determine the explicit form of $u_2(x)$ and plot the functions, u_0, $u_0 + \epsilon u_1$, and $u = u_0 + \epsilon u_1 + \epsilon^2 u_2$ for the value $\epsilon = 0.01$ and for $0 \leq x \leq \pi/2$.
2. Consider the equation

$$u'' + \epsilon u = 0, \qquad u(0) = 1, \qquad u'(0) = 0,$$

and compare the perturbation expansion $u = u_0 + \epsilon u_1 + \cdots$ with the exact solution.

3. Compute u_0 and u_1 for the perturbation expansion $u = u_0 + \epsilon u_1 + \cdots$ of the equation

$$u'' - (1 + \epsilon x)u = 0, \qquad u(0) = u'(0) = 1.$$

Compare with the exact solution e^x of the equation with $\epsilon = 0$.

3.22 THE NONLINEAR SPRING

As a further example of this versatile method, consider the equation for a nonlinear spring,

$$u'' + u + u^3 = 0, \qquad u(0) = c, \qquad u'(0) = 0, \tag{22.1}$$

where c, the initial distension, is small. To put this equation in a form similar to that encountered in the foregoing section, we write $u = cv$. Then v satisfies the equation

$$v'' + v + c^2 v^3 = 0, \qquad v(0) = 1, \qquad v'(0) = 0. \tag{22.2}$$

Replace c^2 by ϵ,

$$v'' + v + \epsilon v^3 = 0, \qquad v(0) = 1, \qquad v'(0) = 0, \tag{22.3}$$

and write

$$v = v_0 + \epsilon v_1 + \epsilon^2 v_2 + \cdots \tag{22.4}$$

By virtue of our assumption concerning c, ϵ may be considered to be a small parameter. Substituting (22.4) in Eq. (22.3), we have

$$\begin{aligned}(v_0'' + \epsilon v_1'' + \epsilon^2 v_2'' + \cdots) + (v_0 + \epsilon v_1 + \epsilon^2 v_2 + \cdots)& \\ + \epsilon(v_0 + \epsilon v_1 + \epsilon^2 v_2 + \cdots)^3 &= 0.\end{aligned} \tag{22.5}$$

Collecting coefficients of powers of ϵ, we obtain

$$(v_0'' + v_0) + \epsilon(v_1'' + v_1 + v_0^3) + \epsilon^2(v_2'' + v_2 + 3v_0^2 v_1) + \cdots = 0. \tag{22.6}$$

Hence, equating coefficients of like powers of ϵ in (22.6) and in the initial conditions, we obtain the equations

$$\begin{aligned} v_0'' + v_0 &= 0, & v_0(0) &= 1, & v_0'(0) &= 0, \\ v_1'' + v_1 + v_0^3 &= 0, & v_1(0) &= 0, & v_1'(0) &= 0, \\ v_2'' + v_2 + 3v_0^2 v_1 &= 0, & v_2(0) &= 0, & v_2'(0) &= 0, \quad \text{etc.}\end{aligned} \tag{22.7}$$

As before, we see that $v_0 = \cos x$. The equation for v_1 is then

$$v_1'' + v_1 + \cos^3 x = 0. \tag{22.8}$$

To obtain the equation for v_2, it is necessary to determine v_1, which we now proceed to do.

3.23 THE EQUATION FOR v_1

We begin with the representation

$$\cos^3 x = \frac{3 \cos x}{4} + \frac{\cos 3x}{4}. \tag{23.1}$$

Thus we must solve the equation

$$v_1'' + v_1 = -\frac{3 \cos x}{4} - \frac{\cos 3x}{4}. \tag{23.2}$$

With a little effort of the type described in Chapter 2, we obtain the particular solution

$$w = -\frac{3x \sin x}{8} + \frac{\cos 3x}{32}. \tag{23.3}$$

To meet the initial conditions, we subtract $(\cos x)/32$. Thus

$$v_1 = -\frac{3x \sin x}{8} + \frac{\cos 3x}{32} - \frac{\cos x}{32}. \tag{23.4}$$

Hence the approximate solution has the form

$$v = \cos x + \epsilon\left(-\frac{\cos x}{32} + \frac{\cos 3x}{32} - \frac{3x \sin x}{8}\right). \tag{23.5}$$

3.24 LEGITIMACY

As we shall presently see, there are certain dangers in the use of perturbation techniques. How can one tell when the expansion is legitimate? The general subject is a very difficult one, and even with the aid of advanced analysis few really powerful results have been found. One useful general property which is not difficult to establish is the following.

Theorem. *Consider the differential equation*

$$y'' + ay' + by = \epsilon g(y, y', x), \qquad y(0) = c_1, \qquad y'(0) = c_2, \tag{24.1}$$

where g is a polynomial in y, y', and x. In a fixed interval $0 \leq x \leq c$, the solution has a power-series expansion in ϵ of the form

$$y = y_0 + \epsilon y_1 + \epsilon^2 y_2 + \cdots, \tag{24.2}$$

where the functions $y_0, y_1, y_2, \ldots,$ are independent of ϵ. The series converges if ϵ is small, $|\epsilon| \leq d$. The radius of convergence d depends on the initial conditions c_1, c_2, the interval length c, and the polynomial g. There is no straightforward way of determining d.

This theorem enables us to see at a glance that perturbation techniques can be used immediately to study such equations as

$$u'' + \epsilon(u^2 - 1)u' + u = 0, \tag{24.3}$$

which is the famous Van der Pol equation, in the interval $0 \leq x \leq 2\pi$.

EXERCISES

1. Carry out perturbation calculations to determine the function $v_2(x)$ in (22.7).

2. Obtain the perturbation solution of

$$u'' + \epsilon(u^2 - 1)u' + u = 0, \qquad u(0) = 1, \qquad u'(0) = 0,$$

$u = \cos x + \epsilon u_1 + \cdots$, up to the determination of u_1.

3. Obtain the perturbation solution of

$$u'' + u + \epsilon u^3 = \cos \omega t,$$

for

$$u(0) = u'(0) = 0,$$

and for

$$\omega \neq 1, u = u_0 + \epsilon u_1 + \cdots,$$

up to the term in u_1.

4. Obtain the perturbation solution $u = u_0 + \epsilon u_1 + \cdots$, of

$$u'' + (1 + \epsilon \cos t)u = 0, \qquad u(0) = 1, \qquad u'(0) = 0,$$

up to the term in u_1.

3.25 DISCUSSION

When we examine the forms of the two perturbation solutions we obtained in Sections 3.21 and 3.23, we observe that both contain terms of the form $x \sin x$. This does not disturb us in the case of the equation

$$u'' + (1 + \epsilon x)u = 0, \tag{25.1}$$

since we do not expect the solution to be periodic in x. It is a bit disturbing in the case of the nonlinear spring, since we do expect a periodic solution there. Of course, we did agree that the perturbation expansion is valid only in a fixed x-interval.

The difficulty resides first of all in the fact that the nonlinearity—and even a change in the constant coefficients—changes the frequency of the

periodic solution. Consider, for example, the equation

$$u'' + (1+\epsilon)^2 u = 0, \qquad u(0) = 1, \qquad u'(0) = 0, \tag{25.2}$$

with the solution $u = \cos(1+\epsilon)x$. The expansion in ϵ is

$$u = \cos x - \epsilon x \sin x + \cdots \tag{25.3}$$

Terms such as $x \sin x$ in a perturbation expansion, which increase arbitrarily in magnitude as the independent variable increases, are called *secular terms*. They were first encountered by mathematicians studying celestial mechanics, and many ingenious techniques have been developed to eliminate them.

For example, returning to Eq. (23.5), we can write this equation in the form

$$v = \left(1 - \frac{\epsilon}{32}\right)\cos x - \frac{3\epsilon x \sin x}{8} + \frac{\epsilon \cos 3x}{32} + \text{higher-order terms.} \tag{25.4}$$

Recalling the Taylor's-series expansion of (25.3), we can write (25.4) as

$$v = \left(1 - \frac{\epsilon}{32}\right)\cos\left(1 + \frac{3\epsilon}{8}\right)x + \frac{\epsilon\cos 3x}{32} + \text{higher-order terms,} \tag{25.5}$$

where the higher-order terms are now different from those in (25.4). The point is that the first-order approximation,

$$v = \left(1 - \frac{\epsilon}{32}\right)\cos\left(1 + \frac{3\epsilon}{8}\right)x + \frac{\epsilon\cos 3x}{32}, \tag{25.6}$$

is now a respectable uniformly bounded function. To make it periodic with period $2\pi/(1 + 3\epsilon/8)$, we can write

$$v \cong \left(1 - \frac{\epsilon}{32}\right)\cos\left(1 + \frac{3\epsilon}{8}\right)x + \frac{\epsilon\cos 3(1 + 3\epsilon/8)x}{32}, \tag{25.7}$$

a change affecting only terms of order ϵ^2 and higher.

3.26 WHEN DO PERIODIC SOLUTIONS EXIST?

We have been blithely proceeding on the assumption that the solution of

$$u'' + u + \epsilon u^3 = 0, \qquad u(0) = 1, \qquad u(0) = 0, \tag{26.1}$$

is indeed periodic in x. Intuitively, we argue as follows: When $\epsilon = 0$ the

solution is periodic. A small value of ϵ should produce a periodic solution close to the original.

The fact that an assertion of this type is not uniformly true has many important biological, engineering, and physical consequences. It is true for the equation of the nonlinear spring introduced in Section 3.22. It is not true for the Van der Pol equation

$$u'' + \epsilon(u^2 - 1)u' + u = 0. \tag{26.2}$$

This equation has a unique periodic solution towards which all other solutions tend, regardless of their initial conditions. This is a very desirable property for an equation which represents the behavior of a multivibrator—a fundamental electronic circuit for World War II radar—or the behavior of the heart.

Obtaining a perturbation expansion for the periodic solution of Eq. (26.2) requires some ingenuity. In general, the study of periodic solutions of nonlinear differential equations requires both advanced and adroit techniques. Even today this area of mathematics is by no means a tamed domain.

3.27 SINGULAR PERTURBATIONS

Suppose that we consider a small mass attached to a spring, which is disturbed from equilibrium position and then left to its own devices. How does the spring-mass system behave? Assume that there is an elastic resistance and a viscous resistance to motion proportional to velocity. The equation describing this motion subject to small displacements is linear and takes the form

$$\epsilon y'' + ky' + cy = 0, \tag{27.1}$$

where y denotes the displacement from equilibrium position. The initial conditions are determined by the position and velocity of the mass at time zero: $y(0) = y_0$, $y'(0) = v_0$.

What happens if we attempt to use a perturbation argument? The resulting equation obtained by setting $\epsilon = 0$ in Eq. (27.1)—that is, the equation which results from disregarding the mass—is

$$ky' + cy = 0; \tag{27.2}$$

it has the general solution

$$y = be^{-ct/k}. \tag{27.3}$$

If we use the initial condition $y(0) = y_0$, we have

$$y = y_0 e^{-ct/k}. \tag{27.4}$$

Equation (27.4) yields the value

$$y'(0) = -\frac{y_0 c}{k}, \tag{27.5}$$

which is usually not equal to v_0. If we use the correct value of $y'(0)$ to determine b, we find a bulge in the determination of $y(0)$.

Which value of $y'(0)$ should be used, and, in general, can we have confidence in an approximate solution obtained from (27.2)? These are very interesting, important, and difficult questions which we do not intend to pursue here. Our aim is to alert the reader to the problem area, possibly to arouse his curiosity, and to provide him with a plenitude of references.

EXERCISES

1. Obtain the explicit analytic solution of

$$\epsilon u'' + u' + u = 0, \qquad u(0) = 1, \qquad u'(0) = 2,$$

for a small ϵ. Draw graphs of the solution in the interval $0 \leq t \leq 1$ for $\epsilon = 0.1, 0.01, 0.001$. Compare these solutions with those obtained from $u' + u = 0$, with an appropriate choice of $b = u(0)$ as far as both $u(t)$ and $u'(t)$ are concerned. What value of b yields a good approximation for large t?

2. Carry out a corresponding analysis for

$$\epsilon u'' + u = 0, \qquad u(0) = 1, \qquad u'(0) = 2.$$

3.28 THE WKB-APPROXIMATION

A basic role in quantum mechanics is played by the one-dimensional Schrodinger equation

$$u'' + \frac{g(x)}{h^2} u = 0, \tag{28.1}$$

where h is a very small constant, called Planck's constant. How does one obtain approximate solutions for an equation of this type? Again, although

the rigorous details would carry us too far afield, we can readily indicate the simple procedure that is used.

It is expected from physical principles that there should be a solution behaving in some fashion like an exponential, $u = e^w$. Consequently, let us employ the following artifice. Write

$$u = \exp\left(\int_0^x v\,dx_1\right) \tag{28.2}$$

so that we can convert (28.1) into the Riccati equation

$$v' + v^2 + \frac{g(x)}{h^2} = 0. \tag{28.3}$$

For small h, which term is larger, v' or v^2? We would expect the square to be larger. Hence let us set

$$v^2 + \frac{g(x)}{h^2} = 0 \tag{28.4}$$

as a first approximation. Thus $v \cong i\sqrt{g(x)}/h$. To obtain a better approximation, we write

$$v = \frac{i\sqrt{g(x)}}{h} + w. \tag{28.5}$$

Substituting (28.5) in Eq. (28.3), we have

$$\frac{ig'}{2h\sqrt{g}} + w' - \frac{g(x)}{h^2} + \frac{2iw\sqrt{g(x)}}{h} + w^2 + \frac{g(x)}{h^2} = 0. \tag{28.6}$$

The resulting equation for w is

$$\frac{ig'(x)}{2h\sqrt{g(x)}} + w' + \frac{2iw\sqrt{g(x)}}{h} + w^2 = 0. \tag{28.7}$$

Let us try to equate the coefficient of $1/h$ to zero to get our next approximation. Thus

$$\frac{ig'}{2\sqrt{g}} + 2iw\sqrt{g} = 0, \tag{28.8}$$

or

$$w = -g'/4g. \tag{28.9}$$

This result is consistent with the approximation made in Eq. (28.7), since w^2 and w' indeed are small compared to a term in $1/h$. Let us then set

(28.10) $$w = -\frac{g'}{4g} + z$$

to obtain the next approximation. At this point, we leave it to the reader to show that we can set

(28.11) $$z = hw_1 + h^2 w_2 + \cdots,$$

where $w_1, w_2, \ldots,$ are independent of h, and proceed to obtain equations for $w_1, w_2, \ldots$

Combining the results obtained so far, we have the classical result

(28.12) $$u \cong \frac{\exp\left[(i/h)\int_0^x \sqrt{g}\,dx_1\right]}{\sqrt[4]{g}}\left[1 + hz_1 + h^2 z_2 + \cdots\right],$$

where the functions $z_1, z_2, \ldots,$ depend only on x, from which many remarkable consequences follow. The series is usually divergent. The first term in this approximation is called the WKB-approximation, despite the fact that it is due to Liouville.

3.29 DISCUSSION

In Section 3.28 we could, of course, have started with a trial solution of the form given (28.12) and then demonstrated that the functions $z_1, z_2, \ldots,$ could be obtained recurrently. The reader then would have wondered, quite justifiably, where we got the inspiration to try this not very obvious expression. We pursued the details in order to give some indication of the techniques that are used to study problems of this type.

The same methods can be used to study the behavior of the solutions to

(29.1) $$u'' + (1 + x)u = 0$$

as $x \to \infty$, or the behavior of the solutions to

(29.2) $$x^4 u'' - u = 0$$

in the neighborhood of $x = 0$. Examples will be given in the following set of miscellaneous exercises, and detailed accounts can be found in the references cited.

MISCELLANEOUS EXERCISES

1. Write $f \ll g$ if

$$f(x) = \sum_{n=0}^{\infty} a_n x^n, \qquad g(x) = \sum_{n=0}^{\infty} b_n x^n, \qquad 0 \leq a_n \leq b_n,$$

and if the series are convergent for $|x| < 1$. Show that $f_1 \ll g_1, f_2 \ll g_2$ imply that $(c_1 f_1 + c_2 f_2) \ll (c_1 g_1 + c_2 g_2)$ for any nonnegative constants c_1 and c_2 and that $f_1 f_2 \ll g_1 g_2$.

2. Given that

$$f(x) = \sum_{n=0}^{\infty} a_n x^n$$

converges for $|x| < 1$, with $a_n \geq 0$, show that for any $b > 1$ we can find a constant k such that $f(x) \ll k/(1 - bx)$.

3. Given that

$$f(x) = \sum_{n=0}^{\infty} a_n x^n, \qquad a_n \geq 0,$$

converges for all x, show that for any $b > 0$ we can find a constant k such that $f(x) \ll k/(1 - bx)$.

4. Show that $f \ll g$ implies that $f' \ll g'$, in Exercise 1.

5. Consider the differential equation

$$u' = p(x)u, \qquad u(0) = 1,$$

where $p(x)$ is a power-series, convergent for $|x| < 1$,

$$p(x) = \sum_{n=0}^{\infty} a_n x^n.$$

Write

$$u(x) = \sum_{n=0}^{\infty} b_n x^n, \qquad v(x) = \sum_{n=0}^{\infty} |b_n| x^n,$$

$$p_1(x) = \sum_{n=0}^{\infty} |a_n| x^n,$$

and let $u_1(x)$ be the solution of $u_1' = p_1(x)u_1$, $u_1(0) = 1$. Show that $v(x) \ll u_1(x)$.

6. Consider the convergence of the power-series expansion of the solution of

$$u_2' = \frac{k}{(1 - bx)u_2}, \qquad u_2(0) = 1,$$

where $k > 0$, $b > 0$, and show that, in view of Exercise 5, the solution of $u' = p(x)u$, $u(0) = 1$, has a power-series expansion convergent for $|x| < 1$.

7. Consider the second-order linear differential equation

$$(*) \qquad u'' = p(x)u' + q(x)u,$$

where p and q have power-series expansions convergent for $|x| < 1$. Show by means of the comparison equation

$$v'' = \frac{k_1 v'}{1 - bx} + \frac{k_2 v}{(1 - bx)^2}$$

that Eq. (*) has two distinct power-series solutions which are convergent for $|x| < 1$. [*Hint:* The equation for v possesses a solution of the form $(1 - bx)^{-a}$, where $a > 0$.]

8. Using the technique of Exercise 7, show that the power-series solutions converge for all x if $p(x)$ and $q(x)$ possess power-series solutions which converge for all x. (In the foregoing exercises we have presented the basic idea of the Cauchy method of majoration, which is one of the most important ways of establishing the convergence of power-series solutions obtained by the techniques presented in this chapter.)

9. Consider the function

$$u(x) = \int_0^\infty \frac{e^{-t}\,dt}{1 + xt},$$

defined for all $x \geq 0$. Show that

$$u'(x) = -\int_0^\infty \frac{te^{-t}\,dt}{(1 + xt)^2}, \qquad u''(x) = \int_0^\infty \frac{2t^2 e^{-t}\,dt}{(1 + xt)^3},$$

and that $u(x)$ satisfies a linear differential equation of the second order.

10. Show that there is a power-series solution for the differential equation obtained in Exercise 9 of the form $S(x) = 1 - x + 2!x^2 - 3!x^3 + \cdots$, divergent for $x \neq 0$.

11. Show that

$$\frac{1}{1 + xt} = 1 - xt + (x^2t^2) + \cdots + \frac{(-1)^n(xt)^n}{1 + xt}.$$

12. Hence show that, in Exercise 9,

$$u(x) = \int_0^\infty e^{-t}\left[1 - xt + x^2t^2 + \cdots + \frac{(-1)^n(xt)^n}{(1 + xt)}\right] dt$$
$$= 1 - x + 2!x^2 \cdots + (-1)^n x^n \int_0^\infty \frac{t^n e^{-t}\,dt}{(1 + xt)}.$$

(Recall that $\int_0^\infty t^k e^{-t}\,dt = k!$ for $k = 1, 2, \ldots$)

13. From the result of Exercise 12 show that

$$|u(x) - [1 - x + 2!x^2 + \cdots + (-1)^{n+1}(n-1)!x^{n-1}]|$$
$$= x^n \int_0^\infty \frac{t^n e^{-t}\,dt}{(1 + xt)} \le x^n \int_0^\infty t^n e^{-t}\,dt = n!x^n.$$

14. Show that for a fixed value of x the function $n!x^n$ first decreases as n assumes the values $1, 2, \ldots,$ and then increases. For what value, or values, of n does it attain its minimum value?

15. If $x = 0.1$, what is the best estimate we can obtain for $u(0.1)$ using the method outlined in the previous exercises?

16. Similarly, use the divergent series $S(x)$ to obtain estimates for $u(0.01)$, $u(0.001)$. (The purpose of the foregoing exercises is to illustrate what we meant when we said that divergent power series were not as useless as one might imagine for obtaining numerical values. The series $S(x)$ is an example of an asymptotic series. The theory of asymptotic series was created by Stieltjes and Poincaré.)

17. Consider the linear differential equation

$$v'' - v = 0, \qquad v(0) = 1, \qquad v'(0) = 0,$$

with the solution $v = (e^x + e^{-x})/2$. Show that the equation $u' + u^2 = 1$, $u(0) = 0$, possesses the solution $u = (e^x - e^{-x})/(e^x + e^{-x})$.

18. Set $u = x/(1 + w)$ in Exercise 17. Show that w satisfies the Riccati equation

$$-xw' = -x^2 + w + w^2.$$

Set $w = a_1x^2/(1 + w_1)$ and show that w_1 satisfies an equation of similar type with different coefficients.

19. Continuing in this fashion, obtain the formal continued-fraction expansion

$$\frac{e^x - e^{-x}}{e^x + e^{-x}} = \cfrac{x}{1 + \cfrac{a_1x^2}{1 + a_2x^2 + \cdots}}$$

and determine the form of the coefficients $a_1, a_2, \ldots$

20. Compare the numerical values of $(e^x - e^{-x})/(e^x + e^{-x})$ at $x = 0.1, 0.5$, and 1, with those obtained by using the second and third convergents,

$$C_2 = \frac{x}{1 + a_1x^2}, \qquad C_3 = \frac{x}{1 + a_1x^2/(1 + a_2x^2)}.$$

Compare with the values obtained by using the first two and the first three terms in the power-series expansion of the function

$$\frac{e^x - e^{-x}}{e^x + e^{-x}}.$$

21. Obtain a similar continued-fraction expansion for $\log [1/(1 - x)]$, starting with the fact that $u(x)$ satisfies the equations

$$u'(x) = 1/(1 - x), \qquad [(1 - x)u']' = 0.$$

22. Obtain a similar continued-fraction expansion for the function $w'(x)/w(x)$, where w satisfies the equation

$$w'' - xw = 0, \qquad w(0) = 1, \qquad w'(0) = 0.$$

(The technique just presented is due to Laguerre.)

23. Examine the behavior of the solutions of $u'' - x^2u = 0$ as $x \to \infty$ in the following fashion. The function $v = u'/u$ satisfies the equation $v' + v^2 - x^2 = 0$. Write $v = x + v_1$, so that v_1 satisfies the equation $v_1' + 2xv_1 + 1 + v_1^2 = 0$. Now obtain an approximate value of v_1 by equating the sum of the two largest terms to zero. Show that the only choice of v_1 consistent with the condition that v_1 is of smaller order than x as $x \to \infty$ (i.e., $v_1/x \to 0$ as $x \to \infty$) is $v_1 \cong -1/2x$.

24. Hence set $v_1 = -1/2x + v_2$. Obtain the equation for v_2 and show that it can be formally satisfied by a series of the form

$$v_2 = \frac{a_2}{x^2} + \frac{a_3}{x^3} + \cdots$$

Determine the coefficients.

25. Show that we could have obtained a formal series expansion of the form

$$u = \frac{e^{\pm x^2/2}}{x^{1/2}}\left[1 + \frac{b_1}{x} + \frac{b_2}{x^2} + \cdots\right],$$

by using the second-order differential equation and calculating the coefficients $b_1, b_2, \ldots$, recursively. Where does the term $e^{\pm x^2/2}/x^{1/2}$ arise?

26. Carry out a similar analysis for $u'' \pm x^4u = 0$ and $u'' \pm x^3u = 0$.

27. What type of behavior do we expect for the solutions of $x^3u'' - u = 0$ as $x \to 0$? Carry out the analysis in two ways, first by means of the foregoing techniques and secondly by setting $x = 1/x_1$.

BIBLIOGRAPHY AND COMMENTS

Section 3.1. The use of power-series expansions in the solution of differential equations was initiated by Newton.

Section 3.2. The convergence of the power-series expansion of a function $f(x)$ depends essentially on the behavior of $f(x)$ as a function of a complex variable x. It was the study of the convergence of a particular series, that obtained from the solution of the Kepler equation, that led Cauchy to the discovery of this remarkable fact. The Kepler equation also stimulated Lagrange to discover the Lagrange expansion.

Section 3.3. For an elementary discussion of asymptotic series, see:

BELLMAN, R., *Perturbation Techniques in Mathematics, Physics and Engineering*, Holt, Rinehart & Winston, New York, 1964.

The Lagrange expansion is also discussed there.

Section 3.4. Exercise 8. For a discussion of the Newton-Raphson technique, see:

BELLMAN, R., and R. KALABA, *Quasilinearization and Nonlinear Boundary Value Problems*, American Elsevier, New York, 1965.

Section 3.5. For a discussion of some important transformations which accelerate the convergence of an infinite series, see:

SHANKS, D., "Nonlinear transformations of divergent and slowly convergent sequences," *J. Math. Phys.* **34**, 1955, pp. 1–42.

Section 3.6. For a treatment of existence and uniqueness of solutions of differential equations and of a number of other advanced areas in the theory of differential equations, see:

BELLMAN, R., *Stability Theory of Differential Equations*, Dover Publications, New York, 1969.

CODDINGTON, E. A., and N. LEVINSON, *Theory of Ordinary Differential Equations*, McGraw-Hill, New York, 1955.

Section 3.9. For a detailed analysis of the convergence of power-series expansions of solutions for differential equations, see:

INCE, E. L., *Ordinary Differential Equations*, Dover Publications, New York, 1944.

Section 3.17. See the book by Ince referred to above.

Section 3.18. Exercise 7. For a detailed treatment of the hypergeometric function, see:

WHITTAKER, E. T., and G. N. WATSON, *A Course of Modern Analysis*, Cambridge University Press, London, 1935, Chapters X and XIV.

Section 3.20. See the book by Bellman, mentioned above in connection with Section 3.3, for a number of references and for many further results.

Section 3.24. An important book in the domain of solutions of nonlinear differential equations with applications to circuit analysis is the following:

MINORSKY, L., *Nonlinear Oscillations*, D. Van Nostrand, Toronto, 1962.

Section 3.26. A rigorous treatment of the existence of periodic solutions requires a certain amount of geometric, or topological, reasoning. See, for example:

LEFSCHETZ, S., *Differential Equations, Geometric Theory*, Interscience, New York, 1957.

Section 3.27. See

ERDELYI, A., "Singular Perturbations," *Bull. Amer. Math. Soc.* **68,** 1962, pp. 420–424,

where other references may be found.

Section 3.28. The importance of this approximation is explained in books on quantum mechanics.

CHAPTER 4

The Numerical Solution of Differential Equations

4.1 INTRODUCTION

The reader may still remember the shock, and perhaps even moral indignation, he felt upon first discovering that there existed simple elementary functions which could not be integrated explicitly in terms of other elementary functions. Particularly important examples are the integrals

$$(1.1) \qquad \int_0^x e^{-x_1^2}\,dx_1 \qquad \text{and} \qquad \int_1^x \frac{e^{-x_1}\,dx_1}{x_1},$$

which occur in numerous scientific investigations. We must reluctantly accept the fact that these expressions represent new functions which are to be studied in their own right. To bring them within the family of "domesticated" functions such as e^x, $\sin x$, and $\log x$, we either provide tables of values or simple algorithms for finding these values. We sometimes forget that the only reason that we consider the aforementioned functions to be tame is that we can readily obtain their numerical values from tables which others have prepared. Thus, when we say that the solution of

$$(1.2) \qquad u'' + u' - 2u = 0, \qquad u(0) = 2, \qquad u'(0) = -1,$$

is

$$(1.3) \qquad u(t) = e^t + e^{-2t},$$

we are implicitly asserting that it is easier to study the properties of the function $u(t)$ using the representation (1.3) than directly in terms of the defining equation (1.2). This assumption is certainly valid from the standpoint of qualitative investigations, but it is true numerically only if we possess the required tables for the exponential function.

What happens when we encounter a function $u(t)$ defined as the solution of an unfamiliar equation such as

$$\frac{du}{dt} = u^2 + t, \qquad u(0) = 1? \tag{1.4}$$

The first thing that we might do is thumb through the pages of a book devoted to the solutions of various types of differential equations and try to find an equivalent equation whose solution is known in terms of existing tabulated functions. These volumes are analogs to the familiar tables of integrals encountered in elementary calculus. In many cases, however, it is quicker to proceed with a direct computational approach. This is especially true even if some explicit analytic solution exists, when a digital computer is available. Frequently, as we hinted at here and there in the preceding chapter, we can also deduce a number of the required analytic properties directly from the equations.

The objective of this chapter is to discuss the simple ideas behind a number of methods currently used in the numerical solution of differential equations, and to illustrate these methods by means of some examples. Then we shall discuss the use of a digital computer in connection with numerical solutions, and give some examples. Some typical flow charts and computer programs involving FORTRAN will be displayed.

It must be emphasized that we are not primarily interested in the most efficient numerical algorithms to employ in connection with the numerical solution of differential equations. These are properly a part of a separate course in numerical analysis. Our objective here is to set the cultural stage for the further education of the reader in the use of the digital computer (and soon the hybrid computer) for fun and profit in the fields of science and mathematics.

4.2 A PURSUIT PROCESS

In Chapter 1 we posed the problem of determining the path of a dog chasing a rabbit. Our assumption is that the rabbit runs in a straight line and the direction of the dog's motion is always pointed towards the rabbit (see Fig. 4.1). A differential equation for the curve traced out by the dog is readily obtained by translating the foregoing statement into the statement that the tangent to the curve, traced out by the dog's motion, at any point P (at any time t) intersects the x-axis at R, which is the position of the rabbit at time t. The situation is graphically illustrated in Fig. 4.2.

We can now obtain a differential equation, which, with a little dexterity, can be explicitly solved. Such dexterity, however, is of no particular importance since small changes in the formulation of the problem readily pro-

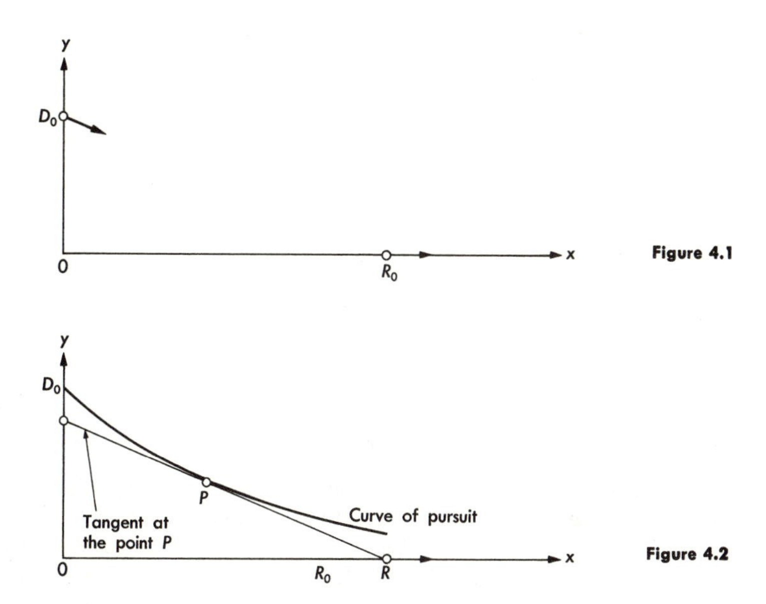

Figure 4.1

Figure 4.2

duce equations which cannot be explicitly resolved in terms of elementary functions. Consequently, in this book we have avoided ad hoc techniques as much as possible.

How then do we determine the form of the curve of pursuit, and obtain such information as the time required for the dog to catch the rabbit?

4.3 A DISCRETE PURSUIT PROCESS

Let us make the problem concrete by assigning some numerical values. Suppose that, at the start of the chase, the rabbit is at +100 ft on the x-axis and the dog is on the y-axis at +50 ft. Let the rabbit travel at 40 ft/sec and the dog at 60 ft/sec. Considering that 88 ft/sec = 60 mi/hr, these are reasonable speeds. To obtain an approximate numerical solution, we begin by making some reasonable approximations concerning the nature of the pursuit. Instead of assuming that the dog changes his direction continuously, we shall suppose that he starts out by running in a straight line for a second in the direction of the rabbit's initial position, changes his direction at the end of one second upon seeing where the rabbit is, continues in a straight line in this new direction for another second, again observes the position of the rabbit, and so on.

It is quite easy to study this new simplified path of pursuit, graphically. Begin with the dog and rabbit in their initial positions, D_0 and R_0, which we connect by means of a ruler. According to our simplifying hypothesis,

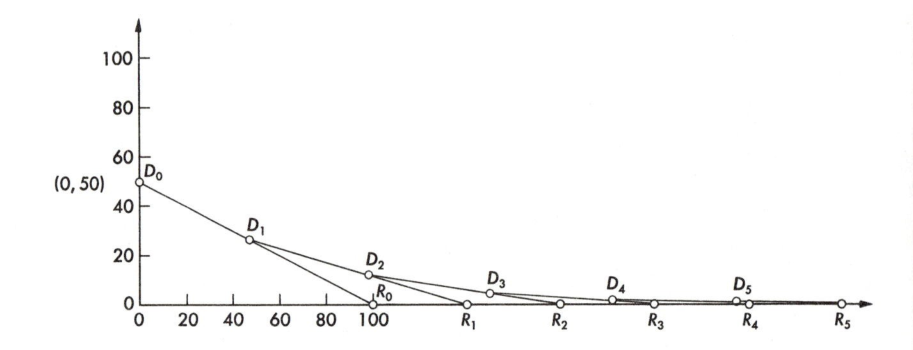

Figure 4.3

the dog continues along this path for a distance of 60 ft during the first second. His new position is D_1. Meanwhile, the rabbit has traversed a distance of 40 ft and is at the position R_1. We now repeat this process, using the points D_1 and R_1 as initial positions. We then obtain two new points D_2 and R_2 (see Fig. 4.3). From this crude approach we estimate that the time of capture is about 6 sec in the original chase. The polygonal curve traced out by the dog consists of the line segments D_0D_1, D_1D_2, D_2D_3, D_3D_4, D_4D_5, D_5D_6, all of equal length, together with a final segment which is not precisely determined because of our condition that the dog runs in one direction for a second at a time. This leads to "overshoot."

To diminish the amount of overshoot, and thus obtain a more accurate capture time, let us suppose that we allow the dog to pursue any fixed direction for only a quarter of a second at a time. A graphical illustration of his path is shown in Fig. 4.4. We see that the "time of capture" according to this new approximation is between 5.25 and 5.50 sec.

It is clear that we should be able to obtain in this way more and more accurate estimates of the time of capture, and closer and closer approximations to the path traced out by the pursuer. We can indeed obtain such accurate estimates, and the result is the substance of a mathematical formulation we shall discuss later. If we want great accuracy, say to one part in a million, we can permit the dog to travel in a fixed direction for only a millionth, or a ten-millionth of a second. There is no conceptual difficulty involved in thinking in these terms, which allow the dog to observe the actual position of the rabbit at shorter time intervals and thus guide his course more accurately. But how can we carry out the arithmetic involved? The answer is that it is too much of a task for human computers, but that it is a task ideally suited to the simple, but persevering, talents of an electronic computer. We will discuss these matters below.

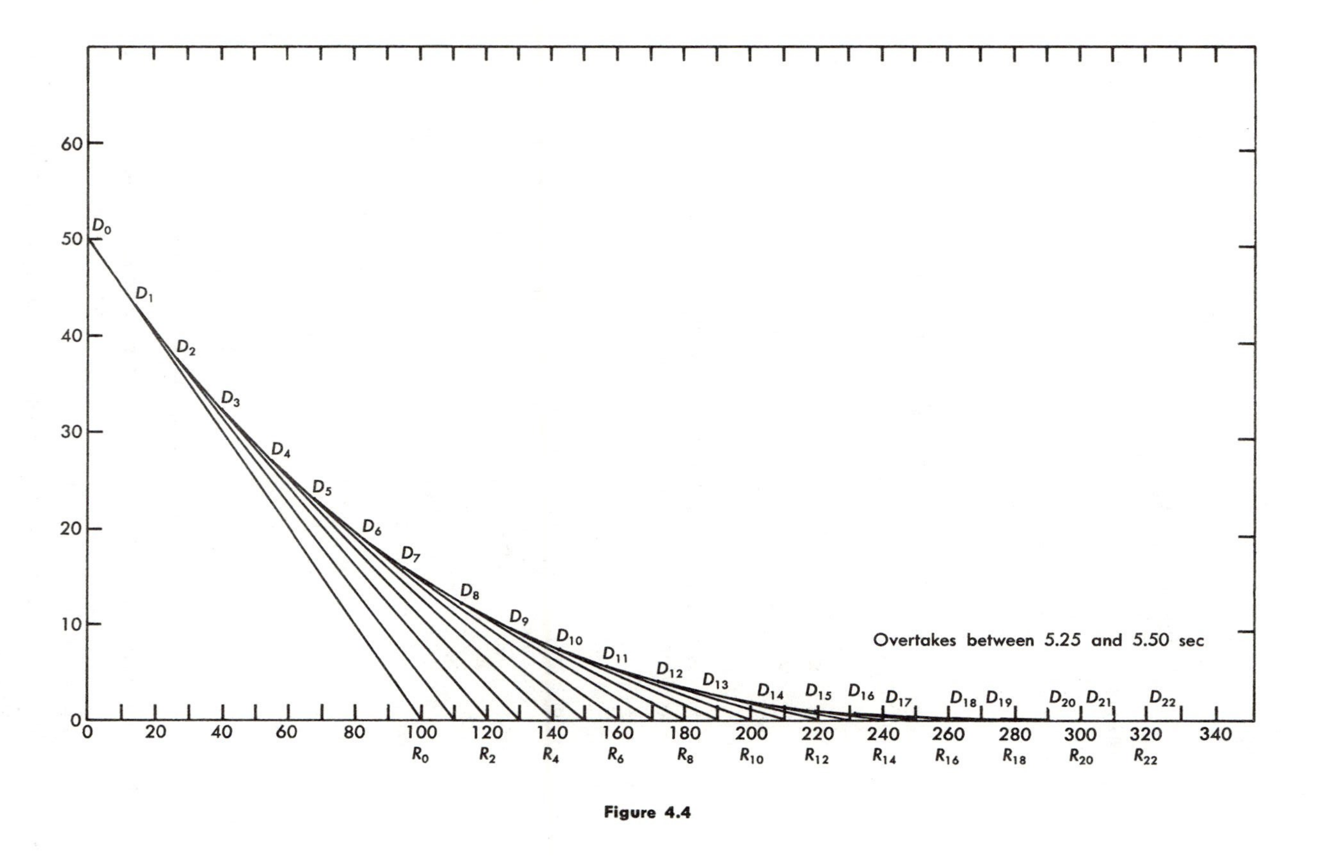
Overtakes between 5.25 and 5.50 sec
D0
D1
D2
D3
D4
D5
D6
D7
D8
D9
D10
D11
D12
D13
D14
D15
D16
D17
D18
D19
D20
D21
D22
R0
R2
R4
R6
R8
R10
R12
R14
R16
R18
R20
R22
0
20
40
60
80
100
120
140
160
180
200
220
240
260
280
300
320
340
0
10
20
30
40
50
60

Figure 4.4

EXERCISES

1. Let (x, y) denote the position of the dog at any time t. Show that the position of the rabbit at the same time is given by $(x - y/y', 0)$. Here $y' = dy/dx$.
2. Show that the condition that the velocity of the dog is n times that of the rabbit leads to the relation

$$\frac{d^2x/dy^2}{[1 + (dx/dy)^2]^{1/2}} = -\frac{1}{ny}.$$

3. Consider the above equation (Exercise 2) a differential equation of the first order for the quantity $u = dx/dy$,

$$\frac{du/dy}{(1 + u^2)^{1/2}} = -\frac{1}{ny},$$

   and integrate both sides between the initial value of y and the current value of y.
4. Solve the resulting equation for $u = dx/dy$ and integrate once again to find x as a function of y. Distinguish between $n = 1$ and $n > 1$.
5. What is the capture time for $n > 1$ as a function of the initial positions, a and b, and so on?
6. Consider a graphical solution in terms of the foregoing type of approximation for the case where the dog is at the center of a circle of radius a and the rabbit is traversing the circumference of this circle. Will the dog ever catch the rabbit if his velocity is equal to that of the rabbit? If it is less?
7. Is there a pursuit policy that will enable the dog to catch the rabbit regardless of the ratio of their velocities?

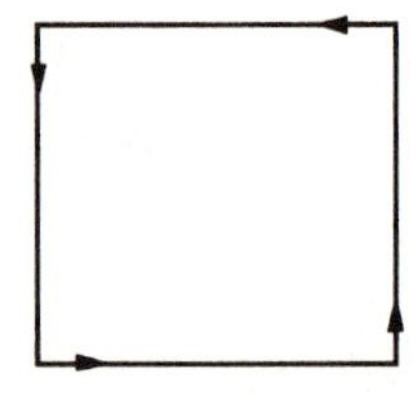

Figure 4.5

8. Consider three dogs at the vertices of an equilateral triangle who chase each other in sequence in the fashion described above. Obtain some graphical approximations to their motions.
9. Consider the same type of problem for four dogs at the vertices of a square (Fig. 4.5).

4.4 DIFFERENCE APPROXIMATIONS

What is rather remarkable about the foregoing presentation is that we could obtain an approximate numerical solution without writing a single differential equation, indeed without writing a single equation of any type. A number of important problems can be attacked in this direct fashion; in particular, electronic computers can often be used in this way. The art of carrying out calculations of this type is often called "simulation." There are many textbooks in this area to which we shall refer the interested reader.

In general, however, when dealing with the question of obtaining numerical answers, we must start with an equation which represents a mathematical transcription of the behavior of the underlying system. The next step is to replace the original equation by a simpler equation which involves only arithmetic operations—which is to say, addition, subtraction, multiplication, and division. In this way, we can obtain an approximate solution by means of a paper-and-pencil calculation, with a slide rule, a desk computer, or, finally, an electronic computer. For the moment, we shall restrict our attention to the use of a desk computer.

In order to indicate the fundamental idea guiding our thoughts in its simplest form, let us consider the equation

$$u' = -u, \qquad u(0) = 1. \tag{4.1}$$

What we want to do is to find an analytic equivalent of the graphical approach used in the study of the pursuit problem. We know the exact solution of Eq. (4.1), $u = e^{-t}$, which makes it unusually easy to investigate the accuracy of our numerical technique.

To obtain an approximate solution to (4.1), we interpret the equation as a set of directions for telling us how the function $u(t)$ changes continuously with t as a function of its current value. Indeed, this is how we derived equations of this nature in Chapter 1. Let us now employ the same trick used above in deriving an approximate solution of the pursuit process. We suppose that u increases at a fixed rate for an initial time of duration Δ, then at another fixed rate determined by the new value of u at time Δ for a second time interval of length Δ, and so on. In order to be flexible, let us for the moment take the time interval Δ to be a parameter which we can subsequently determine in some convenient fashion. Then, graphically we have Fig. 4.6.

We obtain the point P_1 by assuming that $u' = -1$ in the interval $[0, \Delta]$. This value is the approximation obtained from Eq. (4.1), taking cognizance of the fact that $u(0) = 1$. Then, since the path is a straight line, we see that

$$u(\Delta) = 1 - \Delta. \tag{4.2}$$

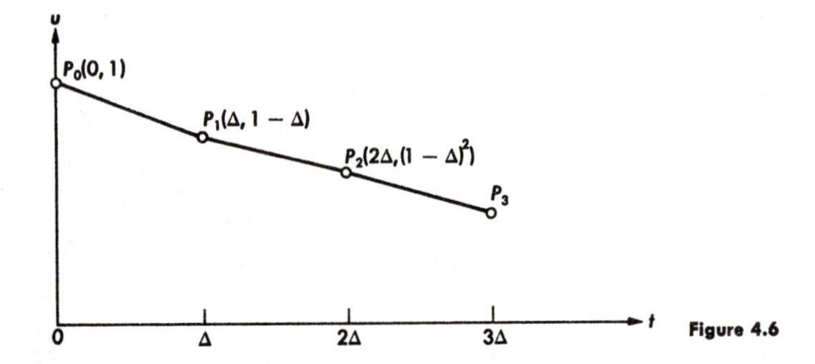

Figure 4.6

For the interval [Δ, 2Δ] we write as our approximation

$$u' = -(1 - \Delta), \tag{4.3}$$

since $u(\Delta) = 1 - \Delta$. Using the initial condition $u(\Delta) = 1 - \Delta$, we readily obtain

$$u(2\Delta) = (1 - \Delta)^2. \tag{4.4}$$

The assumption that the derivative u' is a *constant* in each of these intervals is equivalent to the approximation of the actual function $u(t)$ by a straight line in each of these intervals. In other words, we are using the approximation

$$u'(t) \cong \frac{u(t + \Delta) - u(t)}{\Delta}. \tag{4.5}$$

The Taylor's-series expansion of $u(t + \Delta)$ about the point t,

$$u(t + \Delta) = u(t) + \Delta u'(t) + \frac{\Delta^2}{2} u''(t) + \cdots, \tag{4.6}$$

yields the relation

$$\frac{u(t + \Delta) - u(t)}{\Delta} = u'(t) + \frac{\Delta}{2} u''(t) + \cdots \tag{4.7}$$

We see then that our approximation method consists in using the first two terms in the right-hand side of (4.6) to estimate $u(t + \Delta)$. This, of course, is exactly what a straight-line approximation means. We mentioned this idea in connection with the use of the Newton-Raphson approximation, (see Exercise 8, Section 3.4).

Consequently, instead of the differential equation (4.1), we consider the difference approximation

(4.8) $$\frac{u(t+\Delta)-u(t)}{\Delta} = -u(t), \qquad u(0) = 1,$$

where t now assumes only the values $0, \Delta, 2\Delta, \ldots$ The other values of $u(t)$ are determined by straight-line interpolations. In other words, we connect the points $P_0, P_1, P_2, P_3, \ldots,$ by straight line segments in order to get the graph of the approximate solution $u(t)$. Equation (4.8) is called a *difference equation*, and the technique involving its use is called *difference approximation.* It is one of the fundamental techniques in modern numerical analysis, and serves many theoretical purposes as well.

4.5 EXPLICIT SOLUTION OF THE DIFFERENCE EQUATION

From Eq. (4.8) we see that

(5.1) $$u(t+\Delta) = (1-\Delta)u(t), \qquad u(0) = 1.$$

Hence

(5.2) $$\begin{aligned} u(\Delta) &= (1-\Delta), \\ u(2\Delta) &= (1-\Delta)u(\Delta) = (1-\Delta)^2, \\ u(3\Delta) &= (1-\Delta)u(2\Delta) = (1-\Delta)^3, \end{aligned}$$

and we obtain inductively

(5.3) $$u(n\Delta) = (1-\Delta)^n, \qquad \text{for } n = 1, 2, \ldots$$

Table 4.1

Time	$u_1(t)$	$u_2(t)$	e^{-t}
0.0	0.10000000E 01	0.10000000E 01	0.10000000E 01
0.1	0.90000000E 00	0.90438198E 00	0.90483743E 00
0.2	0.80999999E 00	0.81790676E 00	0.81873076E 00
0.3	0.72899999E 00	0.73970013E 00	0.74081823E 00
0.4	0.65609999E 00	0.66897147E 00	0.67032006E 00
0.5	0.59048998E 00	0.60500574E 00	0.60653067E 00
0.6	0.53144099E 00	0.54715629E 00	0.54881164E 00
0.7	0.47829688E-00	0.49483828E-00	0.49658532E-00
0.8	0.43046720E-00	0.44752282E-00	0.44932898E-00
0.9	0.38742048E-00	0.40473158E-00	0.40656967E-00
1.0	0.34867842E-00	0.36603194E-00	0.36787946E-00

First, consider the case where $\Delta = 0.1$. Call the function obtained with this value of Δ, $u_1(t)$, where t assumes only the values $t = 0.1, 0.2, 0.3$, etc. Consider next the case where $\Delta = 0.01$. Call the function thus obtained $u_2(t)$, where t assumes only the values $t = 0.01, 0.02, 0.03$, etc.

In Table 4.1 we compare the functions $u_1(t)$ and $u_2(t)$ with the exact solution e^{-t}, using the values at $t = 0, 0.1, 0.2, 0.3, \ldots, 1.0$, as convenient check points. We see that the decrease in the size of the Δ-interval from 0.1 to 0.01 naturally improves the accuracy of the approximation. For the benefit of readers not familiar with the notation involving E in Table 4.1, we point out that E stands for exponent and is followed by the appropriate power of ten. Thus, for example,

$$0.10000000\text{E}01 = 0.10000000 \times 10^1 = 1.00000000,$$
$$0.90483743\text{E}00 = 0.90483743 \times 10^0 = 0.90483743.$$

EXERCISES

---

1. We suspect that as Δ is taken smaller we get more and more accurate results. Show that, in the limit as Δ approaches zero, we obtain exact results. [*Hint:* $e^x = \lim_{n\to\infty} (1 + x/n)^n$.]

2. Obtain an estimate for the error committed in using $u_1(t)$ and $u_2(t)$ as approximations to $u(t) = e^{-t}$.

3. Use the approximation formula

$$\frac{u(t+\Delta) - u(t-\Delta)}{2\Delta} = -u(t)$$

in place of Eq. (5.1) for $t = \Delta, 2\Delta, \ldots$, with

$$u(0) = 1, \qquad u(\Delta) = 1 - \Delta + \Delta^2/2.$$

Calculate the approximate solutions obtained for $\Delta = 0.01$ and 0.1, and compare with the numerical results of Table 4.1.

4. Considering

$$u(t+\Delta) - u(t-\Delta) = -2\,\Delta u(t),$$
$$u(0) = 1, \qquad u(\Delta) = 1 - \Delta + \Delta^2/2,$$

as a linear difference equation, obtain the explicit analytic solution and use this for the numerical calculations of Exercise 3.

4.6 DISCUSSION

The method employed in the last section can be applied with relatively minor modifications, which we shall discuss subsequently, to obtain the numerical solutions of many types of differential equations which are quite obdurate analytically. Consider, for example, an equation such as

$$u' = u^2 + t, \qquad u(0) = 1. \tag{6.1}$$

Before attempting to use various computational techniques, it is essential that we have some assurance that a solution exists, and we prefer the situation where there is a unique solution. Although a rigorous discussion of these matters is not very difficult, it would distract us from the principal objective of this chapter. We shall discuss questions of existence and uniqueness briefly in Chapter 6 in order to give the reader some flavor of the fundamental techniques that are used.

Prior to stating a convenient criterion for the existence and uniqueness of the solution of a differential equation of the form

$$u' = g(u, t), \qquad u(0) = c, \tag{6.2}$$

let us give some simple examples which give us some idea of what we can reasonably expect a theorem in this area to tell us.

4.7 LACK OF UNIQUENESS

Consider the simple equation

$$u' = u^{1/2}, \qquad u(0) = 1. \tag{7.1}$$

It has the solution $u = \frac{1}{4}(2 + t)^2$. Generally, the equation

$$u' = u^{1/2}, \qquad u(0) = k, \tag{7.2}$$

possesses the solution

$$u = \tfrac{1}{4}(t + 2\sqrt{k})^2. \tag{7.3}$$

Observe, however, what happens when $k = 0$. The function $u = t^2/4$ is certainly a solution. But so also is $u = 0$.

An example of this type produces a certain uneasiness. How do we know that the same kind of phenomenon does not occur in other equations, equations which may not possess explicit solutions to guide us. In elementary algebra, we meet the gremlin of "extraneous solution." It is certainly reasonable to expect to encounter its big brothers in connection with the far more complicated equations we are now studying.

EXERCISES

1. State what is wrong with the following argument: Starting with $u' = u^{1/2}$, $u(0) = 0$, we have $u'/u^{1/2} = 1$. Integrate both sides between 0 and t; we get $2u^{1/2} = t + c$. Since $u = 0$ at $t = 0$, we must have $c = 0$. Hence $u = t^2/4$, which is the only solution.
2. State what is wrong with the following argument: Set $u = v^2$. The resulting equation is $v' = \frac{1}{2}$. Therefore $v = t/2$, and $u = t^2/4$.
3. Given that $0 \leq a < 1$, $c > 0$, show that $u' = u^a$, $u(0) = c$, possesses one and only one positive solution for $t \geq 0$.

4.8 LACK OF CONTINUITY FOR LARGE t

Consider the two differential equations

$$u' = 1 - u^2, \qquad u(0) = 0, \tag{8.1a}$$

$$u' = 1 + u^2, \qquad u(0) = 0, \tag{8.1b}$$

which are so similar in structure. The first has the solution

$$u = \frac{e^{2t} - 1}{e^{2t} + 1}, \tag{8.2}$$

perfectly well behaved for $t \geq 0$. The second has the solution (more properly, at this point, *a* solution)

$$u = \tan t, \tag{8.3}$$

which is continuous for $0 \leq t < \pi/2$, but which becomes unbounded as t approaches $\pi/2$. We cannot continue the solution past this point $t = \pi/2$.

This is an effect which we had not previously encountered in connection with initial-value problems. Indeed, no such phenomenon occurs for linear equations with constant coefficients subject to initial conditions. How do we know when there exists a solution valid for all $t \geq 0$, or when there is a critical value at which the solution becomes infinite? These are very difficult matters, and not much is known about them. Nor do we ever expect to possess any uniformly applicable techniques. The field of nonlinear differential equations is a jungle, with hard-won outposts of civilization existing only here and there.

EXERCISES

1. Study the solution of $u' = 1 - u^2$, $u(0) = a$, in the three cases $-1 < a < 1$, $a > 1$, $a < -1$, by writing $u'/(1 - u^2) = 1$ and integrating both sides with respect to t. Describe the behavior of the solutions as $t \to \infty$.
2. What is the situation if $a = \pm 1$?

4.9 A USEFUL EXISTENCE AND UNIQUENESS THEOREM

In view of the examples given in the previous sections, we can expect to have little more than the following theorem.

Theorem. *Consider the equation*

$$u' = g(u, t), \qquad u(0) = c. \tag{9.1}$$

If $g(u, t)$ is a polynomial in u and t, then there exists a solution in some interval $0 \leq t \leq t_0$. The positive number t_0 depends on the form of the polynomial $g(u, t)$ and the initial value. This solution is unique.

As we mentioned before, we know of no systematic techniques for finding the length of the interval of existence. Sometimes we can use analytic techniques to gain valuable information, sometimes we rely on physical intuition, and sometimes we are forced to try a computational solution to see what happens.

The foregoing result enables us to look at an equation such as (8.1a), (8.1b), or (6.1) and to know immediately that these equations possess unique solutions in some t-interval $[0, t_0]$. On the other hand, (7.1) escapes the foregoing criterion and, as we noted, actually possesses a multiple solution.

There are slight extensions of the foregoing theorem, but it is reasonably representative of the kind of all-purpose result that one can expect.

4.10 A SIMPLE SAMPLE NUMERICAL APPROACH

Armed with the preceding theorem, let us see how we would go about calculating the numerical values of the solution of

$$u' = u^2 + t, \qquad u(0) = 0.1, \tag{10.1}$$

in the interval $0 \leq t \leq 1$. Since this equation has a right-hand side that is less than $u^2 + 1$, we suspect that the solution will exist in the interval [0, 1], since the solution of the comparison equation

$$u' = u^2 + 1, \qquad u(0) = 0.1, \tag{10.2}$$

namely $u = \tan t$, exists in this interval. Comparison arguments of this type are extremely useful in the theory of differential equations, but we shall make no incursion into this interesting bag of tricks here.

Let us agree initially to divide the interval [0, 1] into ten equal parts. Thus the Δ of Section 4.5 is now 0.1. The quantity Δ is called the *grid size*. In place of (10.1), we write the difference approximation

$$\frac{u(t + \Delta) - u(t)}{\Delta} = u(t)^2 + t, \qquad u(0) = 0.1, \tag{10.3}$$

where t now assumes only the values $0, \Delta, 2\Delta, \ldots, 9\Delta$. Rewriting (10.3), we have

$$u(t + \Delta) = u(t) + \Delta[u(t)^2 + t], \qquad u(0) = 0.1. \tag{10.4}$$

Thus

$$\begin{aligned} u(\Delta) &= u(0) + \Delta[u(0)^2 + 0], \qquad u(0) = 1, \\ u(2\Delta) &= u(\Delta) + \Delta[u(\Delta)^2 + \Delta], \\ u(3\Delta) &= u(2\Delta) + \Delta[u(2\Delta)^2 + 2\Delta], \\ &\vdots \\ u(1) = u(10\Delta) &= u(9\Delta) + \Delta[u(9\Delta)^2 + 9\Delta]. \end{aligned} \tag{10.5}$$

Hence purely by means of arithmetic operations, we can calculate first $u(\Delta)$, then $u(2\Delta)$ knowing the first value, and so on. The values thus obtained are, of course, approximations to the values of the solution of Eq. (10.1),

Table 4.2 ($\Delta = 0.1$)

t	(1) t	(2) $u(t)$	(3) $u(t)^2$	(4) $u(t)^2 + t$	(5) $\Delta[u(t)^2 + t]$	(6) $u(t) + \Delta[u(t)^2 + t]$
			$(2)^2$	(1) + (3)	0.1 (4)	(2) + (5)
0	0.0	0.100000	0.010000	0.010000	0.001000	0.101000
Δ	0.1	0.101000	0.010201	0.110201	0.011020	0.112020
2Δ	0.2	0.112020	0.012548	0.212548	0.021255	0.133275
3Δ	0.3	0.133275	0.017762	0.317762	0.031776	0.165051
4Δ	0.4	0.165051	0.027242	0.427242	0.042724	0.207775

but we expect these approximations to become better and better as Δ is chosen smaller and smaller.

Let us actually carry out the calculations, assuming either that the reader is a rapid arithmetician or, what is far more sensible, that he possesses a desk calculator. It is convenient to arrange the calculation as in Table 4.2. (Recall that $\Delta = 0.1$ in this calculation.)

EXERCISES

1. Fill in the remaining five lines of Table 4.2, corresponding to the t-values 0.5, 0.6, 0.7, 0.8, 0.9, i.e., 5Δ, 6Δ, 7Δ, 8Δ, 9Δ.
2. Use linear interpolation to obtain an approximate value of the solution of Eq. (10.1) at $t = 0.35$.
3. Use linear extrapolation to obtain an approximate value of this solution at $t = 1.05$.
4. Find the first six terms of the power-series expansion of the solution of Eq. (10.1) and compare the values for $t = 0, 0.1, 0.2, \ldots, 0.9, 1.0$, with those obtained in Table 4.2.

4.11 GREATER ACCURACY

If we wanted a more accurate determination of the solution of Eq. (10.1), we would expect that we could obtain it by decreasing the value of Δ. Suppose, for example, in the above calculation using (10.3) we take $\Delta = 0.05$. Then the first six lines of the corresponding table are as in Table 4.3. Observe that the values of u corresponding to $t = 0.1$ and 0.2 are slightly different.

Table 4.3 ($\Delta = 0.05$)

t	(1) t	(2) $u(t)$	(3) $u(t)^2$	(4) $u(t)^2 + t$	(5) $\Delta[u(t)^2 + t]$	(6) $u(t) + \Delta[u(t)^2 + t]$
			$(2)^2$	(1) + (3)	0.05 (4)	(2) + (5)
0	0.00	0.100000	0.010000	0.010000	0.000500	0.100050
Δ	0.05	0.100050	0.010010	0.060010	0.003001	0.103051
2Δ	0.10	0.103051	0.010620	0.110620	0.005531	0.108582
3Δ	0.15	0.108582	0.011790	0.161790	0.008090	0.116672
4Δ	0.20	0.116672	0.013612	0.213612	0.010681	0.127353
5Δ	0.25	0.127353	0.016219	0.266219	0.013311	0.140664

Why not take a grid size of 0.01, 0.001, or even 0.000001? What stops us from obtaining arbitrary accuracy in this simple fashion? One part of the answer is very simple: arithmetic. There is nothing conceptually difficult about the foregoing idea, but the execution of it is a different matter. It cannot be overemphasized that it is extremely hard to do arithmetic—correctly, that is.

Assuming, then, that we must get along with the results produced with a Δ of reasonable size, how accurate are the results obtained? Is there a theory which can be used to predict the size of errors to be expected for a given Δ? Such a theory of error does exist, but is quite complicated. We shall touch on it briefly below, but refer the reader to more advanced books for more thorough discussions. A point that can be made now, however, is that a procedure much used in practice is that of comparison of computations made with different values of the grid size, for example Δ and $\Delta/2$. If the two tables of values coincide to a certain number of decimal digits, we accept those digits as correct. If the number of digits of agreement is not sufficient, we decrease Δ and see if results are better.

The point to emphasize here is that in numerical computation it is essential to analyze the computed data carefully. Blind acceptance of such data can sometimes mean that although pages of six or seven digit numbers are printed, none is significant!

EXERCISES

1. Fill out the remaining lines of Table 4.3 corresponding to the grid size $\Delta = 0.05$, and graph the two approximate solutions.
2. Compare this new approximation with the approximation obtained by the use of a truncated power-series expansion.

4.12 VALIDITY OF DIFFERENCE APPROXIMATION

We have been operating on the assumption that, for Δ taken sufficiently small, the values obtained from the difference approximation yield arbitrarily accurate estimates of the true solution. This is indeed the case under the hypothesis of the theorem stated in Section 4.9. As we shall point out, however, there is much more to the art of numerical solution of differential equations than this comforting result might indicate. At the end of this chapter we shall provide some references to textbooks on numerical analysis where these matters are discussed in some detail.

4.13 HOW DO WE OBTAIN GREATER ACCURACY?

We have deliberately assigned as exercises to the reader the onerous task of filling out the tables in Sections 4.10 and 4.11 in order to make the reader aware of what we mean when we say that arithmetic is hard and, indeed, painful. Furthermore, it is time-consuming. Using the foregoing method, a calculation based on the use of a grid size of 0.01 would take about ten times as long to carry out as one using 0.1 as grid size; a calculation based on a grid size of 0.000001 would take about one hundred thousand times as long. This is too long considering the order of importance of this problem.

Let us see if we can employ some of our mathematical ingenuity to save our energies for more interesting occupations than sustained arithmetic calculation. Instead of the approximation

$$u'(t) \cong \frac{u(t+\Delta) - u(t)}{\Delta}, \tag{13.1}$$

let us use the approximation

$$u'(t) \cong \frac{u(t+\Delta) - u(t-\Delta)}{2\Delta}. \tag{13.2}$$

Since

$$u(t+\Delta) = u(t) + \Delta u'(t) + \frac{\Delta^2}{2}u''(t) + \frac{\Delta^3}{6}u'''(t) + \cdots, \tag{13.3}$$

$$u(t-\Delta) = u(t) - \Delta u'(t) + \frac{\Delta^2}{2}u''(t) - \frac{\Delta^3}{6}u'''(t) + \cdots,$$

we see that (13.2) yields the relation

$$\frac{u(t+\Delta) - u(t-\Delta)}{2\Delta} = u'(t) + \frac{\Delta^2}{6}u'''(t) + \cdots \tag{13.4}$$

On the other hand, (13.1) leads to

$$\frac{u(t+\Delta) - u(t)}{\Delta} = u'(t) + \frac{\Delta u''(t)}{2} + \cdots \tag{13.5}$$

Consequently, assuming that u', u'', u''' are all well-behaved functions in the t-interval of interest (which is to say, they are functions which do not vary in value in any drastic manner), we see that (13.2) yields a significantly better approximation for small Δ than (13.1). Let us then attempt to find a numerical solution to Eq. (10.1) using the difference approximation

$$\frac{u(t + \Delta) - u(t - \Delta)}{2\Delta} = u(t)^2 + t, \qquad u(0) = 0.1. \tag{13.6}$$

Now t assumes the values $\Delta, 2\Delta, 3\Delta, \ldots$ Thus,

$$\begin{aligned} \frac{u(2\Delta) - u(0)}{2\Delta} &= u(\Delta)^2 + \Delta, \\ \frac{u(3\Delta) - u(\Delta)}{2\Delta} &= u(2\Delta)^2 + 2\Delta, \quad \text{etc.} \end{aligned} \tag{13.7}$$

We have the value of $u(0)$ given to us, $u(0) = 0.1$. The recurrence relations (13.7) can be used to calculate $u(2\Delta), u(3\Delta), \ldots$, provided that we can obtain the value of $u(\Delta)$ in some way.

Let us calculate $u(\Delta)$ by using a Taylor-series expansion around the point $t = 0$. We have

$$u(\Delta) = u(0) + \Delta u'(0) + \frac{\Delta^2}{2} u''(0) + \frac{\Delta^3}{6} u'''(0) + \cdots \tag{13.8}$$

Since the differential equation is

$$u' = u^2 + t, \qquad u(0) = 0.1, \tag{13.9}$$

we see that $u'(0) = 0.01$. Differentiating the equation above, we have

$$u'' = 2uu' + 1. \tag{13.10}$$

Hence $u''(0) = 1.002$. Differentiating again, we obtain

$$u''' = 2u'^2 + 2uu'', \tag{13.11}$$

whence

$$u'''(0) = 2(0.0001) + 2(0.1)(1.002) = 0.2006.$$

Thus

$$\begin{aligned} u(\Delta) &\cong 0.1 + (0.01)\Delta + (0.501)\Delta^2 + \frac{0.2006}{6}\Delta^3 \\ &\cong 0.1 + (0.01)\Delta + 0.5\Delta^2. \end{aligned} \tag{13.12}$$

We suspect that this value will be accurate enough for our purposes in view of the order of accuracy of (13.4). Having obtained $u(\Delta)$, we can proceed to calculate $u(2\Delta), u(3\Delta), u(4\Delta), \ldots$, recurrently using (13.6).

EXERCISE

1. Take $\Delta = 0.1$, and use (13.6) to calculate the approximate values of the solution of Eq. (13.9) at the t-values 0.1, 0.2, . . . , 0.9, 1.0. How do these values compare with those found previously, and with values obtained using the power-series expansion?

4.14 NUMERICAL STABILITY

Having seen how a difference approximation such as (13.6) yields a more accurate approximation than (10.3), we might imagine that we could do even better by using more refined approximations.

Methods of this type are important, but they must be carefully examined, since extraneous solutions can cause trouble, as we wish now to indicate. Suppose that we desire to solve

$$u' = -u, \qquad u(0) = 1, \tag{14.1}$$

numerically (overlooking the fact that we have an explicit analytic solution), and that we use the difference approximation

$$\frac{u(t + \Delta) - u(t - \Delta)}{2\Delta} = -u(t), \qquad u(0) = 1, \tag{14.2}$$

valid for $t = \Delta, 2\Delta, 3\Delta, \ldots$ If we set

$$u(k\Delta) = u_k, \qquad t = k\Delta, \tag{14.3}$$

$k = 0, 1, 2, \ldots$, then (14.2) yields

$$\frac{u_{k+1} - u_{k-1}}{2\Delta} = -u_k, \tag{14.4}$$

or

$$u_{k+1} - u_{k-1} + 2\,\Delta u_k = 0. \tag{14.5}$$

The initial conditions are $u_0 = 1$, $u_1 = e^{-\Delta} \cong 1 - \Delta + \Delta^2/2$. Equation (14.5) is a linear difference equation, which, as we know, has the particular solutions r_1^k and r_2^k, where r_1, r_2 are the roots of

$$r^2 + 2\,\Delta r - 1 = 0. \tag{14.6}$$

Hence

$$r = \frac{-2\Delta \pm \sqrt{4\Delta^2 + 4}}{2}. \tag{14.7}$$

Thus $r_1 \cong 1 - \Delta$, $r_2 \cong -1 - \Delta$ for small Δ. It follows that the general solution of (14.5) is given by

$$u_k = c_1(1 - \Delta)^k + c_2(-1 - \Delta)^k. \tag{14.8}$$

Let us see what values of c_1 and c_2 are required to fit the initial conditions $u_0 = 1$, $u_1 = 1 - \Delta + \Delta^2/2$. We obtain the linear algebraic equations

$$\begin{aligned} c_1 + c_2 &= 1, \\ (1 - \Delta)c_1 + (-1 - \Delta)c_2 &= 1 - \Delta + \Delta^2/2. \end{aligned} \tag{14.9}$$

An easy calculation yields

$$c_1 = 1 + \Delta^2/4, \qquad c_2 = -\Delta^2/4. \tag{14.10}$$

Thus

$$u_k = \left(1 + \frac{\Delta^2}{4}\right)(1 - \Delta)^k - \frac{\Delta^2}{4}(-1 - \Delta)^k. \tag{14.11}$$

Hence, if $k = t/\Delta$, i.e., $\Delta = t/k$, we have

$$\begin{aligned} u_k &= \left(1 + \frac{\Delta^2}{4}\right)\left(1 - \frac{t}{k}\right)^k - \frac{\Delta^2}{4}\left(-1 - \frac{t}{k}\right)^k \\ &= \left(1 + \frac{\Delta^2}{4}\right)\left(1 - \frac{t}{k}\right)^k - \frac{\Delta^2}{4}(-1)^k\left(1 + \frac{t}{k}\right)^k \\ &\cong \left(1 + \frac{\Delta^2}{4}\right)e^{-t} - \frac{\Delta^2}{4}(-1)^k e^t. \end{aligned} \tag{14.12}$$

The exact solution is e^{-t}. We see that the term $(\Delta^2/4)e^{-t}$ contributes only a small percentage of error if Δ is small, as we hoped. But what about the term $(\Delta^2/4)(-1)^k e^t$? If the t-interval is small, this term still yields a small percentage of error. If, however, the integration is carried out to large values of t, say $t = 10$, then the percentage error is

$$\frac{100|\Delta^2/4|e^t}{e^{-t}} = 100\left|\frac{\Delta^2}{4}\right|e^{20}. \tag{14.13}$$

Since

$$e^{10} = 10^{10\log_{10}e} \cong 10^{10(0.4343)} \cong 10^{4.343}, \tag{14.14}$$

we see that the percentage of error is

$$100\left|\frac{\Delta^2}{4}\right|10^{8.686} = \left|\frac{\Delta^2}{4}\right|10^{10.686}. \tag{14.15}$$

To keep (14.15) small, Δ would have to be chosen very small indeed, e.g., $\Delta = 10^{-6}$. This means that the computation would take much too long a time. Think of the number of arithmetic operations involved in going from $t = 0$ to $t = 10$ with a step size of 10^{-6}!

In Table 4.4 we give the results of a numerical integration of Eq. (14.1) by the method of this section with $\Delta = 0.1$, for $0 \leq t \leq 9$. The table lists the computed values of u, the exact solution e^{-t}, the error $u(t) - e^{-t}$ and the relative error $(u(t) - e^{-t})/e^{-t}$. For the sake of brevity, many of the computed values are not printed. Comparing these values with $u_1(t)$ as given in Table 4.1, we see that for small t the method here is much more accurate, for the same Δ, than the one given there. However, at around $t = 1.5$ a strange thing happens: the relative error, which up to this time

Table 4.4 $\Delta = 0.1$, $u' = -u$, $u(0) = 1$

Time	u	e^{-t}	Error	Relative error
0.0	0.10000000E 01	0.10000000E 01	0.0	0.0
0.1	0.90500003E 00	0.90483743E 00	0.16260147E–03	0.17970240E–03
0.2	0.81900007E 00	0.81873083E 00	0.26923418E–03	0.32884325E–03
0.3	0.74120009E 00	0.74081832E 00	0.38176775E–03	0.51533245E–03
0.4	0.67076015E 00	0.67032015E 00	0.44000149E–03	0.65640477E–03
0.5	0.60704815E 00	0.60653079E 00	0.51736832E–03	0.85299578E–03
0.6	0.54935062E 00	0.54881173E 00	0.53888559E–03	0.98191318E–03
1.5	0.22371525E 00	0.22313035E 00	0.58490038E–03	0.26213394E–02
1.6	0.20240414E 00	0.20189661E 00	0.50753355E–03	0.25138289E–02
1.7	0.18323439E 00	0.18268371E 00	0.55068731E–03	0.30144302E–02
1.8	0.16575724E 00	0.16529900E 00	0.45824051E–03	0.27721915E–02
1.9	0.15008295E 00	0.14956880E 00	0.51414967E–03	0.34375463E–02
2.0	0.13574064E 00	0.13533539E 00	0.40525198E–03	0.29944272E–02
4.0	0.18103573E–01	0.18315673E–01	−0.21209940E–03	−0.11580210E–01
4.1	0.17055154E–01	0.16572699E–01	0.48245490E–03	0.29111423E–01
4.2	0.14692541E–01	0.14995608E–01	−0.30306727E–03	−0.20210400E–01
4.3	0.14116645E–01	0.13568584E–01	0.54806098E–03	0.40391907E–01
4.4	0.11869211E–01	0.12277368E–01	−0.40815771E–03	−0.33244722E–01
4.5	0.11742800E–01	0.11109017E–01	0.63378364E–03	0.57051279E–01
4.6	0.95206499E–02	0.10051861E–01	−0.53121150E–03	−0.52847076E–01
4.7	0.98386705E–02	0.90952963E–02	0.74337423E–03	0.81731677E–01
4.8	0.75529143E–02	0.82297623E–02	−0.67684799E–03	−0.82243919E–01
6.6	−0.31042681E–02	0.13603715E–02	−0.44646375E–02	−0.32819252E 01
6.7	0.61946064E–02	0.12309155E–02	0.49636886E–02	0.40325174E 01
6.8	−0.43431856E–02	0.11137782E–02	−0.54569617E–02	−0.48995047E 01
6.9	0.70632398E–02	0.10077888E–02	0.60554482E–02	0.60086479E 01
7.0	−0.57558306E–02	0.91188448E–03	−0.66677146E–02	−0.73120165E 01

had been steadily but slowly increasing, begins to oscillate. There is an overall increase but from one step to the next there may be an increase or a decrease. This is a sign of the onset of *numerical instability*. However, this is not easily detectable from the computed values of u alone. By time $t = 4.0$, the percentage error has increased to 1 or 2%, and is actually fluctuating in sign. Moreover, the instability is now apparent from the values of u, which are seen to be decreasing in a very uneven way. The percentage error now increases very rapidly and by $t = 6.0$ is almost 100%. Finally, the values of u begin to oscillate with increasing amplitude.

How could we have detected the onset of numerical instability for an equation for which the exact solution is not already known? It is clear that uneven or oscillatory values computed for u would be a warning signal. In this connection it is important to have at least a rough idea of the general behavior of the solution—decreasing, increasing, oscillating, or what have you—so that gross deviations of the computed values from this behavior can be noted.

The reader may point out that the difficulty is due to the fact that we calculated $u(\Delta)$ in too crude a fashion. Why not carry the power-series expansion in (14.7) out to a few more terms and ensure that $u(\Delta)$ is accurate to twenty significant figures? This would yield far more accurate values for c_1 and c_2 in (14.10) and thus decrease the magnitude of the coefficient of e^t.

The reader will note that the difficulty in the use of (14.4) arose from the fact that we were integrating over a long time interval, long enough for e^t to overwhelm the small coefficient $\Delta^2/4$. If the interval were small, say [0, 1], this difficulty would not arise. It should be pointed out clearly that no matter what type of finite difference scheme is used, there are bound to be difficulties of one type or another if the interval of integration is inordinately large.

These difficulties can be overcome in one way or another by means of analytic ingenuity. The techniques used usually require a higher level of mathematical training than that we can reasonably expect of the reader at this point.

4.15 ROUND-OFF ERROR

The answer to the reasonable suggestion made at the end of Section 4.14 lies in a phrase that strikes terror into the hearts of even the boldest of numerical analysts: *round-off error*. The key point is that arithmetic performed with numbers is not the same as algebra performed according to mathematical rules. When we write

$$xy = z, \tag{15.1}$$

we mean that the quantities xy and z are equal, absolutely and exactly equal. When x is a ten-digit number and y is a ten-digit number, and we write (15.1) with z a ten-digit number, we mean that the two sides are equal to ten significant numbers. The error committed is due to rounding z off to ten significant figures.

Let us illustrate this most important point as far as it concerns numerical analysis by means of the following simple sample calculation. Suppose that $x = 0.37$, $y = 0.54$ to two significant figures. Then

$$(0.37)(0.54) = 0.1998 = 0.20, \tag{15.2}$$

to two significant figures. The error in an individual calculation may not be large. If, however, we perform millions of such calculations, as we do in solving large systems of differential equations, or in obtaining the solution of a single equation over an interval of considerable length, then the round-off error can accumulate at an alarming rate. The result may be that the final answer contains gross inaccuracies.

Desk calculators operate to ten significant figures as do most electronic computers routinely. Special provision to calculate to twenty significant figures can be made with current computers. This is called "double precision." In many cases, such a procedure circumvents the obstacle of round-off error at the expense of using a considerably greater amount of computing time.

In connection with the use of Eq. (14.5), we see that a round-off error could be committed in the calculation of u_2, due to the addition of $u_0 - 2\,\Delta u_1$. Thus u_2 would be slightly incorrect. This error would introduce the same type of error as did the approximate value of $u(\Delta) = u_1$, and again introduce the extraneous solution $c_3 e^t$, where c_3 is a small quantity. The point is that there is no way of accruing any advantage from a very precise initial condition if round-off immediately introduces an error. The danger involved in using the approximation (14.2) rather than the less accurate

$$\frac{u(t + \Delta) - u(t)}{\Delta} = -u(t), \qquad u(0) = 1, \tag{15.3}$$

lies in the appearance of the unwanted solution of the second-order difference equation (14.5). More complicated difference approximations can introduce worse effects, unless carefully screened ahead of time.

In the solution of higher-order equations such as

$$u'' - 100u = 0, \qquad u(0) = 1, \qquad u'(0) = -10, \tag{15.4}$$

it is necessary to be even more careful about numerical instability. Here the difficulty is intrinsic to the equation itself rather than due solely to the difference method used. For (15.4), the exact solution is $u = e^{-10t}$, but the other characteristic root $+ 10$ introduces an undesirable solution e^{10t}. For small t and small Δ, the coefficient of e^{10t} will be small but as t increases this solution will become dominant.

In any case, we hope we have indicated that great care must be exercised in the use of numerical techniques. Computing is an art with an associated theory. Both must be learned. Fortunately, in the majority of cases involving ordinary differential equations, we can often use standard techniques that have been closely scrutinized and analyzed, and which have worked well in numerous applications. But eternal vigilance is the price of computational accuracy.

4.16 TAKING THE STING OUT OF ARITHMETIC

Suppose that a scientific investigation required the solution of an equation such as

$$u' = u^2 + t, \qquad u(0) = c, \tag{16.1}$$

for $0 \leq t \leq 1$, for each of ten values of c. It would be an onerous chore, but it could be carried out. If the solution were required for one hundred values of c, we would either "enlist" the aid of some students, purchase the help of professional computers, or embark on a mathematical research program that might provide some convenient analytic representation that would simplify the numerical evaluation. If one thousand, or ten thousand such calculations were required, and no analytic relief were in sight, the scientific investigation would either be dropped or carried on in an inadequate fashion.

This was the situation prior to 1950. Then a new and astonishing device became widely available—an electronic computer that could perform arithmetic a million times faster than ever before. At the present time, it is safe to say that the amplification factor is now one billion, and is increasing each year. We shall suppose in the sequel that the reader is familiar with the rudiments of the properties of a digital computer, and with elementary FORTRAN. Although there are by now many languages in use for communicating instructions to a computer, the principles at the level of interest to us here are the same. Consequently, we shall present our programs in FORTRAN. Let us now investigate the use of the "sorcerer's apprentice" to do our mathematical fetching and carrying.

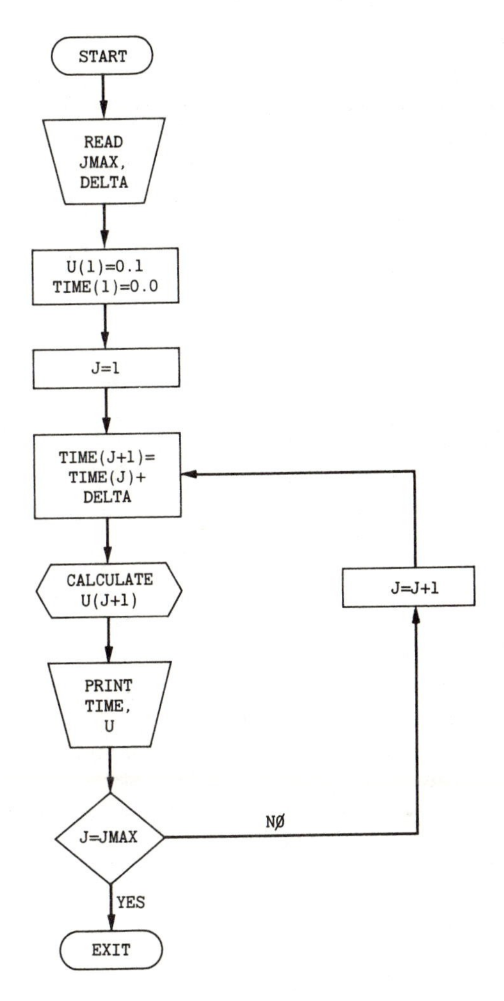

Figure 4.7

4.17 FLOW CHARTS AND FORTRAN PROGRAMS

In order to direct a computer to carry out an arithmetic calculation such as that suggested by the formula

$$\frac{u(t+\Delta)-u(t)}{\Delta} = u(t)^2 + t, \qquad u(0) = 0.1, \tag{17.1}$$

$t = 0, \Delta, \ldots,$ it is first necessary to organize the cycle of calculations. An organization chart of this type is called a *flow chart*. In the case of (17.1), this chart assumes the form shown in Fig. 4.7. The chart must now be translated into explicit instructions for a computer. Thinking in terms of FORTRAN IV, suitable for an IBM 7044 computer, a set of instructions can take the following form

```
      DIMENSIØN U(1001),TIME (1001)
      READ (5,100) JMAX,DELTA
      TIME(1)=0.0
      U(1)=0.1
      DØ 1 J=1,JMAX
      TIME(J+1)=TIME(J)+DELTA
      U(J+1)=DELTA*((U(J)**2)+TIME(J))+U(J)
    1 CØNTINUE
      JMAX1=JMAX+1
      WRITE (6,101) (TIME(K),U(K),K=1,JMAX1)
  100 FØRMAT
  101 FØRMAT
      CALL EXIT
      END
```

The execution time of the program is zero in terms of seconds. Nonetheless, approximately thirty seconds is consumed in each case below, $\Delta = 0.01$ and $\Delta = 0.001$, in connection with various accounting aspects of computer usage. We are, of course, speaking of a fairly old-fashioned computer by contemporary standards.

In Table 4.5, we see the output for $\Delta = 0.01$; in Table 4.6, $\Delta = 0.001$. This is indeed the way to do arithmetic.

Table 4.5

$\Delta = 0.01,\ u' = u^2 + t,\ u(0) = 0.1$

Time	u
0.0	0.10000000E 00
0.1	0.10553362E 00
0.2	0.12128477E 00
0.3	0.14753838E 00
0.4	0.18471760E 00
0.5	0.23346657E 00
0.6	0.29475484E 00
0.7	0.37001942E 00
0.8	0.46137155E 00
0.9	0.57191619E 00
1.0	0.70627313E 00

Table 4.6

$\Delta = 0.001,\ u' = u^2 + t,\ u(0) = 0.1$

Time	u
0.0	0.10000000E 00
0.1	0.10599318E 00
0.2	0.12222593E 00
0.3	0.14899302E 00
0.4	0.18673485E 00
0.5	0.23612235E 00
0.6	0.29816509E 00
0.7	0.37436029E 00
0.8	0.46691157E 00
0.9	0.57907050E 00
1.0	0.71570035E 00

4.18 WHAT VALUE OF Δ TO USE?

How do we know what value of Δ to use? We are caught between two desires. On the one hand, we want a reasonably accurate solution; on the other hand, the time required to carry out the calculation of values of $u(t)$ for t in a fixed interval $[0, t_0]$ is proportional to $1/\Delta$. Experience in computation enables us to look at an equation and estimate a value of Δ. We then use Δ and Δ/2 and compare the answers. If the two tables of values coincide to a specified degree of accuracy, we accept the answer. If there is a disturbing discrepancy when we shift from Δ to Δ/2, we use Δ/2 and Δ/4, or Δ/10 and Δ/20, and see what happens.

As is to be expected, there are theoretical estimates which can be made as to the degree of approximation that is being made for a particular choice of Δ. Usually, they are far too pessimistic.

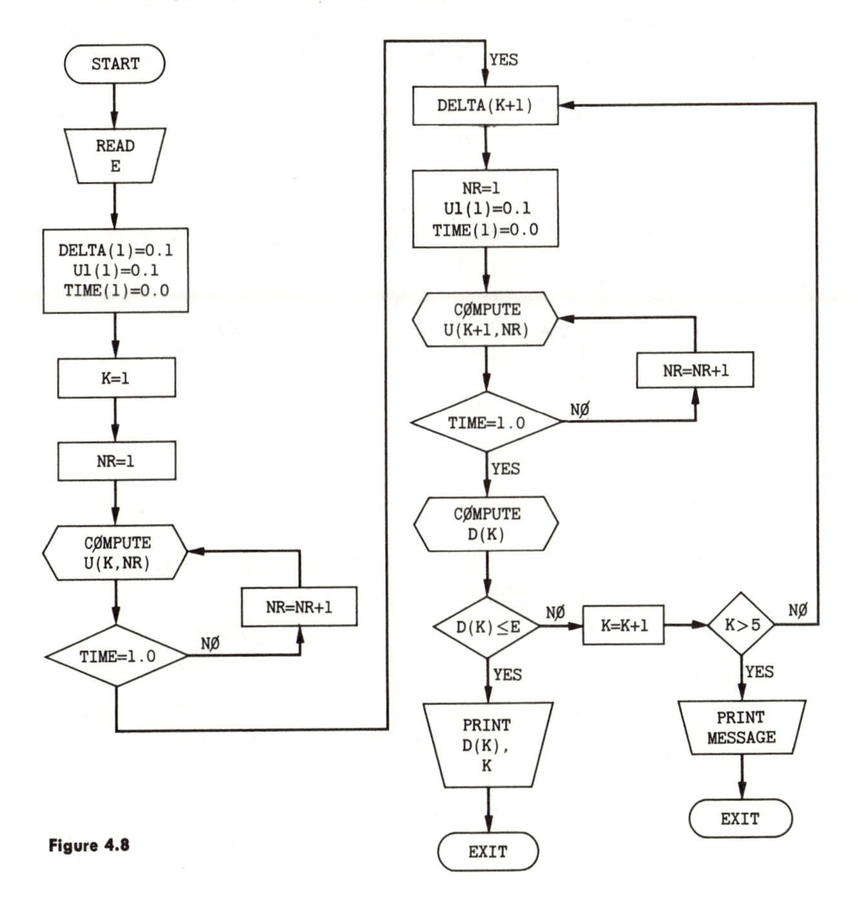

Figure 4.8

Alternatively, if the calculations consume as little time as the foregoing computations do, we can use a simple *adaptive procedure*. Let Δ be chosen, and $u_1(t)$ denote the function determined by this value of Δ. Let $u_2(t)$ denote the function obtained by using $\Delta/2$; and, generally, let $u_k(t)$ denote the function obtained by using $\Delta/2^k$.

Consider the mean-square difference

$$D_k = \sum_{r=1}^{10} [u_k(r) - u_{k+1}(r)]^2. \tag{18.1}$$

Let us calculate D_k for $k = 1, 2, \ldots, N$, and instruct the computer to stop when $D_k \leq \epsilon$, where ϵ is our measure of agreement between two successive calculations.

In this procedure, we do not know in advance how many calculations are required. Since the computation itself determines the number of steps, we use the adjective "adaptive" to describe a procedure of this type. The flow chart now takes the form in Fig. 4.8. A FORTRAN program is given below.

```
C     NOTE-ASSUME IN THIS EXPERIMENT THAT
C            (1) A MAXIMUM OF 5 VALUES OF K ARE TO BE
C                CONSIDERED.
C            (2) INITIAL DELTA=0.1
C            (3) THE 10 TIME POINTS R ARE 0.1, 0.2,
                 . . . , 1.0.
C
      DIMENSION U(6,10), U1(321), TIME(321),
     1 DELTA(6), RTIME(10)
      READ (5,100) E
      DELTA(1)=0.1
      U1(1)=0.1
      TIME(1)=0.0
      K=1
      NR=1
C     SET TIMES AT WHICH EACH FUNCTIONAL VALUE OF U(K)
C     ARE TO BE CONSIDERED
      DO 1 N=1,10
      R=N
      RTIME(NR)=R/10.0
    1 CONTINUE
C     CALCULATE NUMBER OF STEPS REQUIRED, JMAX,
C     USING DELTA(K) TO REACH TIME=1.0
      XTIME=1.0/DELTA(K)
```

```
      JMAX=XTIME
      DØ 3 J=1,JMAX
      TIME(J+1)=TIME(J+1)+DELTA(K)
      U1(J+1)=DELTA(K)*((U1(J)**2)+TIME(J))+U1(J)
      IF (TIME(J+1).EQ.RTIME(NR)) GØ TØ 2
      GØ TØ 3
    2 NR=NR+1
      U(K, NR)=U1(J+1)
    3 CØNTINUE
C     CALCULATE U(K+1,NR)
    4 DELTA(K+1)=DELTA(K)/2.0
      JMAX=JMAX*2
      NR=1
      U1(1)=0.1
      TIME(1)=0.0
      DØ 6 J=1,JMAX
      TIME(J+1)=TIME(J)+DELTA(K+1)
      U1(J+1)=DELTA(K+1)*((U1(J)**2)+TIME(J))+U1(J)
      IF (TIME(J+1).EQ.RTIME(NR)) GØ TØ 5
      GØ TØ 6
    5 NR=NR+1
      U(K+1,NR)=U1(J+1)
    6 CØNTINUE
C     CALCULATE D(K)
      K1=K+1
      SUM=0.0
      DØ 7 N=1,10
      SUM=SUM+(U(K,NR)-U(K1,NR))**2
    7 CØNTINUE
      D(K)=SUM
      IF (D(K).LE.E) GØ TØ 8
      K=K+1
      IF (K.GT.5) GØ TØ 9
      GØ TØ 4
    8 WRITE (6,101) D(K), K
      GØ TØ 10
    9 WRITE (6,102)
  100 FØRMAT
  101 FØRMAT
  102 FØRMAT (20H1K HAS BEEN EXCEEDED)
   10 CALL EXIT
      END
```

4.19 EVALUATION OF FUNCTIONS

Let us consider the equation

$$(19.1) \qquad u' = u^2 + e^t, \qquad u(0) = 1.$$

If we apply the simple numerical method of Section 4.10, we must solve recursively for $u(k\Delta)$, $k = 1, 2, 3, \ldots$, using the difference approximation

$$(19.2) \quad u(t + \Delta) = u(t) + \Delta[u(t)^2 + e^t], \qquad u(0) = 1, \quad t = 0, \Delta, 2\Delta, \ldots.$$

While this method is exactly the same as in Section 4.10, a possible new difficulty is present in that the function which has to be evaluated for $t = 0, \Delta, 2\Delta, \ldots$ is $u(t)^2 + e^t$ instead of $u(t)^2 + t$. If we recall that e^t is a transcendental function which cannot be exactly evaluated in a finite number of simple arithmetic operations, we realize that there may well be significant new difficulties of accuracy and time.

What are the methods available for calculation and use of the values of e^t for $t = 0, \Delta, 2\Delta$, etc.? One, of course, is to read into the storage of the computer the required values as taken from a reliable published table. However, if we are using a value such as $\Delta = 0.005$ and computing $u(t)$ for $0 \leq t \leq 10$, this would mean that 2000 values would have to be stored. For smaller values of Δ, the entire storage capacity of the machine might be exhausted, and in any case it would be very time-consuming to enter all these data into the machine.

A second, and generally preferable, method is to program the machine to compute the required values of e^t when they are needed. In this way, the value of $e^{2\Delta}$, say, can be calculated, used, and discarded, and need never be stored. As a matter of fact, almost any algorithmic language, including FORTRAN, comes with a predesigned program to calculate the simple transcendental functions such as e^t, $\sin t$, $\cos t$, etc. Thus the programmer need only write an instruction containing EXP(T) or some similar expression in order to obtain the value of e^t for any specified t. However, despite the apparent ease with which such values are generated, one must keep in mind that the internal computer program is using some kind of an approximation method* and that a considerable amount of arithmetic may be necessary to compute the approximate value to the specified accuracy. It follows that the evaluation of a function such as $u(t)^2 + e^t$, $t = 0, \Delta, 2\Delta, \ldots$, may be the slowest part of the process of numerical solution of the differential equation. Clearly it is desirable, in order to minimize computation time,

*For example the FORTRAN IV algorithm to compute e^t, as used in the IBM System/360, is based on a continued fraction expansion of e^t.

to use as large a value of Δ as is consistent with obtaining the necessary accuracy of results.

Incidentally, since a function such as e^t is evaluated by some approximation method, it is inevitable that the calculated values are not exact. It follows that at each step of the numerical process a small function evaluation error is committed. This is in addition to the error caused by approximating the differential equation by a difference equation and to the round-off error which occurs in the arithmetic operations.

4.20 USING THE DIFFERENTIAL EQUATION FOR e^t

The function e^t is the unique solution of the differential equation

(20.1)
$$v' = v, \qquad v(0) = 1.$$

Consequently any numerical method for solving (20.1) will generate approximate values of e^t. This fact can be used to circumvent the evaluation of e^t by other methods, as required in the solution of (19.1). In fact, (19.1) is evidently equivalent to the system of coupled differential equations

(20.2)
$$\begin{aligned} u' &= u^2 + v, \qquad & u(0) &= 1, \\ v' &= v, & v(0) &= 1. \end{aligned}$$

Any numerical method for such systems can be used to generate a solution of (20.2). For example, use of the simple difference approximation (4.5) for both u and v yields the pair of difference equations

(20.3)
$$\begin{aligned} u(t + \Delta) &= u(t) + \Delta[u(t)^2 + v(t)], \qquad & u(0) &= 1, \\ v(t + \Delta) &= v(t) + \Delta v(t), & v(0) &= 1. \end{aligned}$$

From these equations, values for $u(t)$ and $v(t)$ at $t = \Delta, 2\Delta, \ldots$ can readily be generated.

By a device such as this, the evaluation of a complicated function is avoided, at the expense of increasing the number of differential equations to be integrated. Whether this represents a reduction in time or storage required will, of course, depend on the particular problem at hand.

EXERCISE

Show that the Mathieu equation $u'' + (a + b \cos t)u = 0$ can be replaced by the system $u'' + (a + bv)u = 0$, $v'' + v = 0$, $v(0) = 1$, $v'(0) = 0$.

4.21 IDENTIFICATION OF SYSTEMS

Once we possess a simple, straightforward approach to the numerical solution of ordinary differential equations subject to initial conditions, we are in an ideal position to begin a frontal attack on one of the most important of all scientific problems: the identification of the structure of a system on the basis of observations and measurements.

In the preceding pages we examined various aspects of the problem of determining the solution given the equation. Let us now follow Jacobi's dictum and invert the problem. How do we determine the equation given the solution?

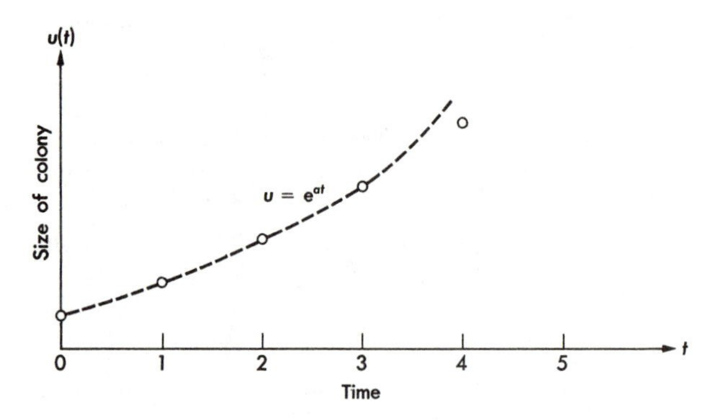

Figure 4.9

Consider, for example, the growth of a colony of bacteria where the size of the colony, measured at hour intervals, yields a graph of the type shown in Fig. 4.9. As we can see, a curve of the form $u = e^{at}$ gives a reasonable fit over the first three hours, but it is clear that the growth levels off after the third hour. Suppose that we want a better fit by using the solution of

$$\frac{du}{dt} = au - bu^2, \qquad u(0) = c, \tag{21.1}$$

where a, b, and c are parameters to be chosen in some expedient fashion.

If we are forced to rely only on our wits with no amplification of our arithmetic talents available, we could proceed in the following fashion. To begin with, let us take $c = u_0$, the population size observed at $t = 0$. To distinguish between the experimentally observed results and the solution of Eq. (21.1), let us use $u(t)$ to denote this solution and u_k to represent the size of the colony observed at the kth hour.

To obtain an estimate for a, we can use the first four values u_0, u_1, u_2, u_3. To obtain an estimate for b, we can keep the colony under observation until it appears to have attained a steady-state population. In this condition, $du/dt = 0$, which means that

$$au - bu^2 = 0. \tag{21.2}$$

Assuming that the population size has not dwindled to zero, we see that the steady-state population size is given by

$$u_\infty = a/b. \tag{21.3}$$

We have thus obtained approximations to the values of a, b, and c.

Although ad hoc techniques of this type can be used in a number of different investigations, requiring more or less ingenuity depending on the nature and accuracy of the measurements, they are neither uniformly reliable nor uniformly effective. For example, it may be quite difficult to measure with accuracy the size of a bacterial population when it is small, and it may be quite time consuming to wait until steady-state growth is attained.

Can we use the power of the digital computer to study these basic questions in a more systematic fashion? It turns out that we can and that we can make great strides toward a coherent theory of formulation, instrumentation, experimentation, and calculation. This is a fascinating tale at whose beginnings we can merely hint in this volume.

4.22 DIRECT SEARCH

Since computers enable us to obtain the numerical solutions of equations such as (21.1) practically instantaneously compared to our usual computation time scale, we can afford to try some very direct and unsophisticated approaches.

Let $u(b, t)$ denote the solution of Eq. (21.1) under the assumption that a and c have been determined according to some method or other. We can try a number of possible values of b, compare the calculated values with the graph in Fig. 4.9, and then choose a best value of b.

With the aid of a modern computer capable of instantaneously displaying the solution curve $u(b, t)$ on an oscilloscope screen, this procedure can be made adaptive. We choose a value b_1, observe the curve $u(b_1, t)$, choose an improved value b_2, observe the new curve $u(b_2, t)$, and so on. With a little practice we can make this procedure into a very efficient one—a nice example of man-machine cooperation.

If the determination of b is part of a larger problem under study with the aid of the computer, or if we want the identification to be done without human intervention, we can proceed in the following manner.

Form the function

$$f(b) = \sum_{k=1}^{5} [u(b, k) - u_k]^2, \tag{22.1}$$

calculate the values at a set of b-points, $b_1, b_2, \ldots, b_M$. Using a straightforward comparison technique, we can have the computer determine the value b_i which yields the smallest of the values $f(b_1), f(b_2), \ldots, f(b_M)$.

A flow chart for this process has the form given in Fig. 4.10, and the FORTRAN program is given below.

```
      READ (5, 100) N
      READ (5, 101)(B(I),I=1,N)
      READ (5, 102)(UØ(K),K=1,5)
      DØ 2 I=1,N
      DØ 1 K=1,5
C     CALCULATE U(I,K) IN THE SAME MANNER AS GIVEN IN
C     THE PRØGRAM ØN PP. 144 AND 145, WHERE THE INDEX
C     I CØRRESPØNDS TØ THE ESTIMATE ØF B THAT IS BEING
C     USED, B(I), AND K CØRRESPØNDS TØ A TIME PØINT AT
C     WHICH AN ØBSERVATION, UØ(K), HAS BEEN MADE.
    1 CØNTINUE
C     CALCULATE F(B)
      SUM=0.0
      DØ 3 K=1,5
      SUM=SUM+(U(I,K)-UØ(K))**2
    3 CØNTINUE
      F(I)=SUM
      IF (I.EQ.1) GØ TØ 5
      IF (F(I).LE.FMIN) GØ TØ 5
      GØ TØ 2
    5 FMIN=F(I)
      IMIN=I
    2 CØNTINUE
      WRITE (6,103) FMIN,IMIN
  100 FØRMAT
  101 FØRMAT
  102 FØRMAT
  103 FØRMAT
      CALL EXIT
      END
```

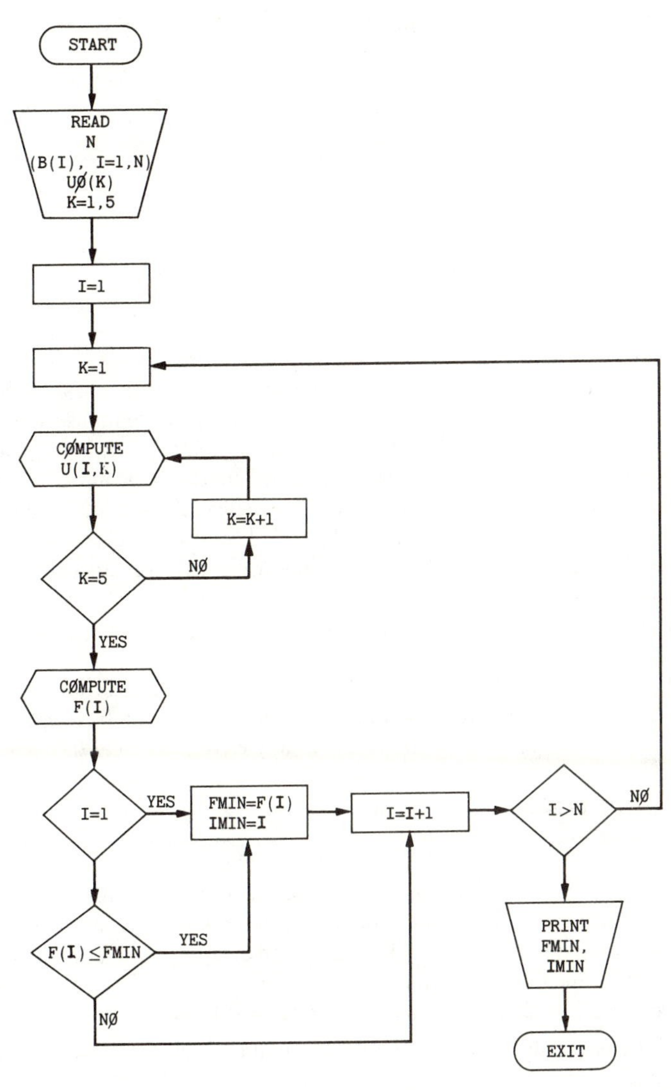

Figure 4.10

4.23 SOPHISTICATED SEARCH

Given the values $f(b_1), f(b_2), \ldots, f(b_M)$, we should be able to do better than merely compare $f(b_1)$ and $f(b_2)$, $f(b_3)$ with the smaller of these numbers, and so on. (See Fig. 4.11.)

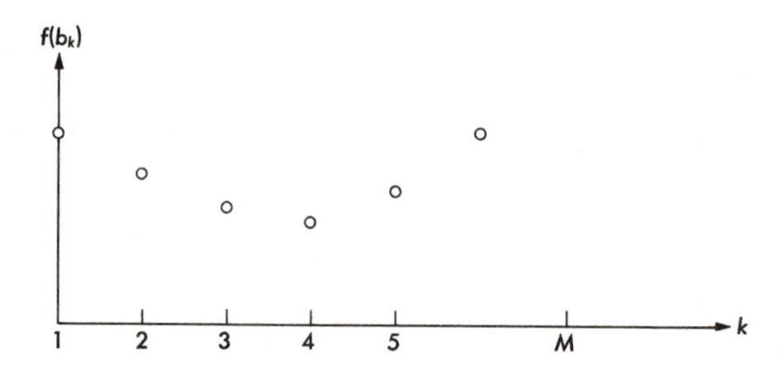

Figure 4.11

For example, if we suppose that there is one and only one value of b_k which yields the smallest value of $f(b)$ for $b = b_1, b_2, \ldots, b_M$, we could use a technique which splits the interval $[1, M]$ in two parts, $[1, M/2]$, $[M/2 + 1, M]$. Let us suppose that M is even. We can then compare the value at $b_{M/2}$ with the value at $b_{(M/2+1)}$. If the value at $b_{M/2}$ is smaller, the minimum lies in $[1, M/2]$; if not, the minimum lies in $[M/2 + 1, M]$. Continuing in this fashion, we can rapidly search for the minimum of $f(b)$ over an extremely large set of points.

A flow chart for a process of this type is given in Fig. 4.12 on pp. 176 and 177.

Search processes over discrete sets of points occupy a central position in the current usage of computers. Combinatorial problems of this type are extraordinarily difficult to solve, and they usually escape classical theories.

MISCELLANEOUS EXERCISES

1. Write $u = 1 + \sum_{k=1}^{\infty} a_k t^k$, and obtain the recurrence relations for the coefficients in the power-series solution of $u' = u^2 + t$, $u(0) = 1$. Write a flow chart and computer program for the calculation of $a_1, a_2, \ldots, a_N$, where N is an input parameter, and for the calculation of $u_N = 1 + \sum_{k=1}^{N} a_k t^k$. Use this program to calculate u_{10}, u_{15}, u_{20} for $t = 0.1, 0.2, \ldots, 0.9, 1.0$.
2. Extend the program obtained in Exercise 1 to an adaptive program where we calculate
$$\epsilon_N = \sum_{k=1}^{10} [u_N(0.k) - u_{N+1}(0.k)]^2$$
for $N = 5, 6, \ldots$, and stop when $\epsilon_N \leq \epsilon$, where ϵ is an input parameter. Run the calculation for $\epsilon = 0.1, 0.01, 0.001, 0.000001$, and examine the results.
3. How many comparisons are required in connection with the search process sketched in Section 4.21 when $M = 1024$? When $M = 2^{20}$? How efficient is this search process as compared to a straightforward pairwise comparison?

4. Design and execute a computer experiment to test the effectiveness of the adaptive program in Section 4.18.

5. Compute solutions of the Michaelis-Menten equation

$$\frac{du}{dt} = -\frac{ku}{u+a}, \qquad u(0) = c,$$

for various values of k, a, c and compare the results.

6. Compute solutions for the equations

$$u'' + \sin u = 0, \qquad u'' + u - \frac{u^3}{6} = 0$$

with $u(0) = 1$, $u'(0) = 0$. Compare the periods of these solutions.

BIBLIOGRAPHY AND COMMENTS

Section 4.1. It is surprisingly difficult to establish the fact that an integral such as

$$\int_0^x e^{-x_1^2}\, dx_1$$

cannot be expressed in terms of the elementary functions of analysis. What is even more surprising is the amount of effort required to make this last statement a precise mathematical statement. For a modern account of the theory which was initiated by Liouville, see:

Hardy, G. H., *The Integration of Functions of a Single Variable*, Cambridge Tracts No. 2, Cambridge University Press, 1958.

Ritt, J. F., *Integration in Finite Terms; Liouville's Theory of Elementary Methods*, Columbia University Press, New York, 1948.

Theories of this type have become of great interest and importance in connection with the study of the abilities and limitations of the digital computer conceived of as an "intelligent machine." Certain parts of this study overlap the domain of mathematical logic.

In asking why we consider $u = e^t + e^{-2t}$ to be a solution of $u'' - u' - 2u = 0$, $u(0) = 2$, $u'(0) = -1$, we are hinting at the fact that the situation may actually be just the reverse in using digital computers. It may be much simpler and more efficient to store the algorithm for calculating $u = e^t + e^{-2t}$, i.e., the differential equation, than to store the functional values of u at a set of grid points, or to calculate u using its explicit analytic form.

The ability of the digital computer to perform rapid arithmetic, together with its (relative) inability to store and retrieve data, has forced us to reexamine critically the concepts of "problem" and "solution."

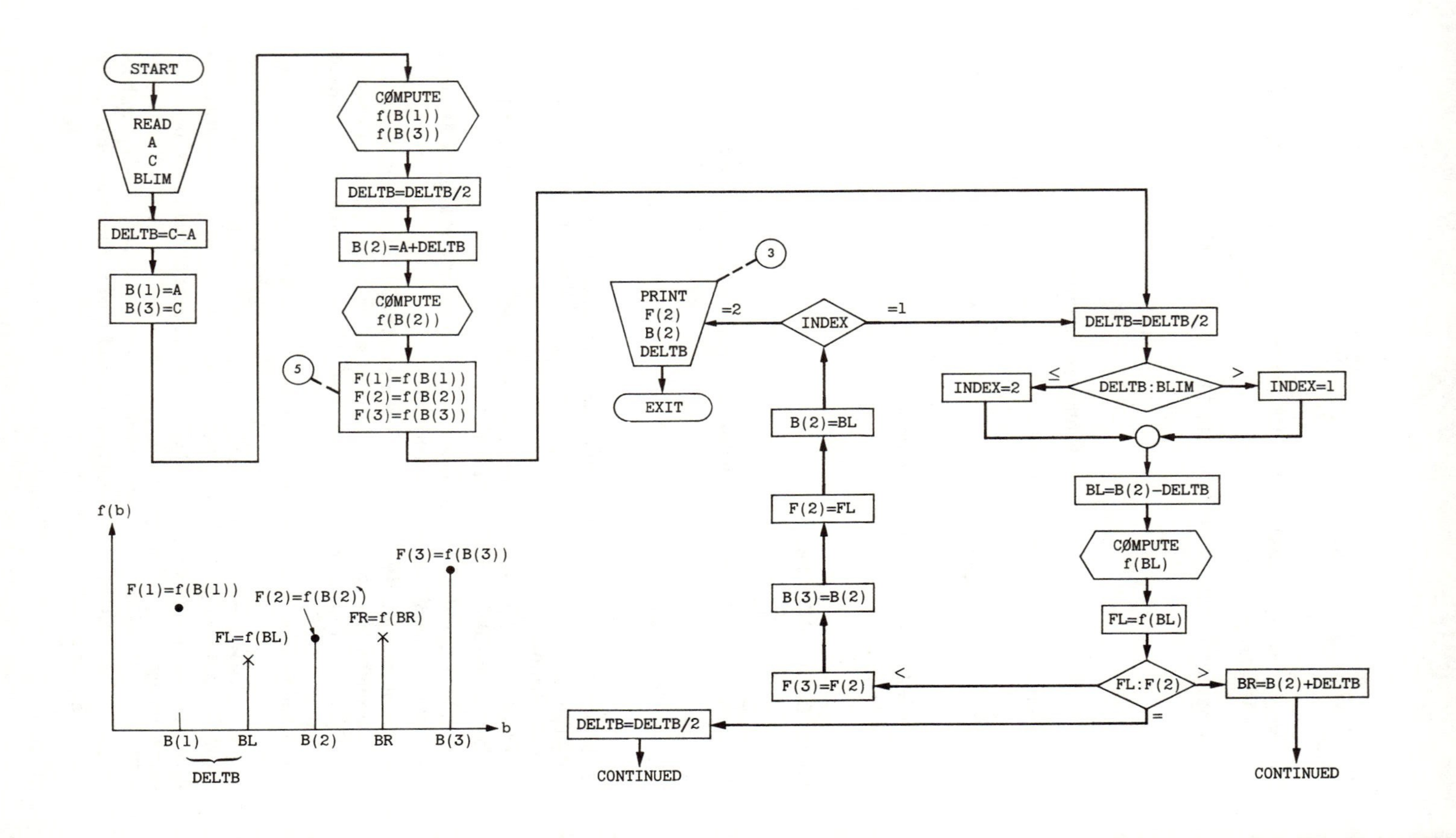
START
READ
A
C
BLIM
DELTB=C-A
B(1)=A
B(3)=C
CØMPUTE
f(B(1))
f(B(3))
DELTB=DELTB/2
B(2)=A+DELTB
CØMPUTE
f(B(2))
F(1)=f(B(1))
F(2)=f(B(2))
F(3)=f(B(3))
5
3
PRINT
F(2)
B(2)
DELTB
EXIT
=2
INDEX
=1
DELTB=DELTB/2
≤
DELTB:BLIM
>
INDEX=2
INDEX=1
B(2)=BL
F(2)=FL
B(3)=B(2)
F(3)=F(2)
BL=B(2)-DELTB
CØMPUTE
f(BL)
FL=f(BL)
<
FL:F(2)
>
=
BR=B(2)+DELTB
DELTB=DELTB/2
CONTINUED
CONTINUED
f(b)
F(1)=f(B(1))
F(2)=f(B(2))
F(3)=f(B(3))
FL=f(BL)
FR=f(BR)
b
B(1)
BL
B(2)
BR
B(3)
DELTB

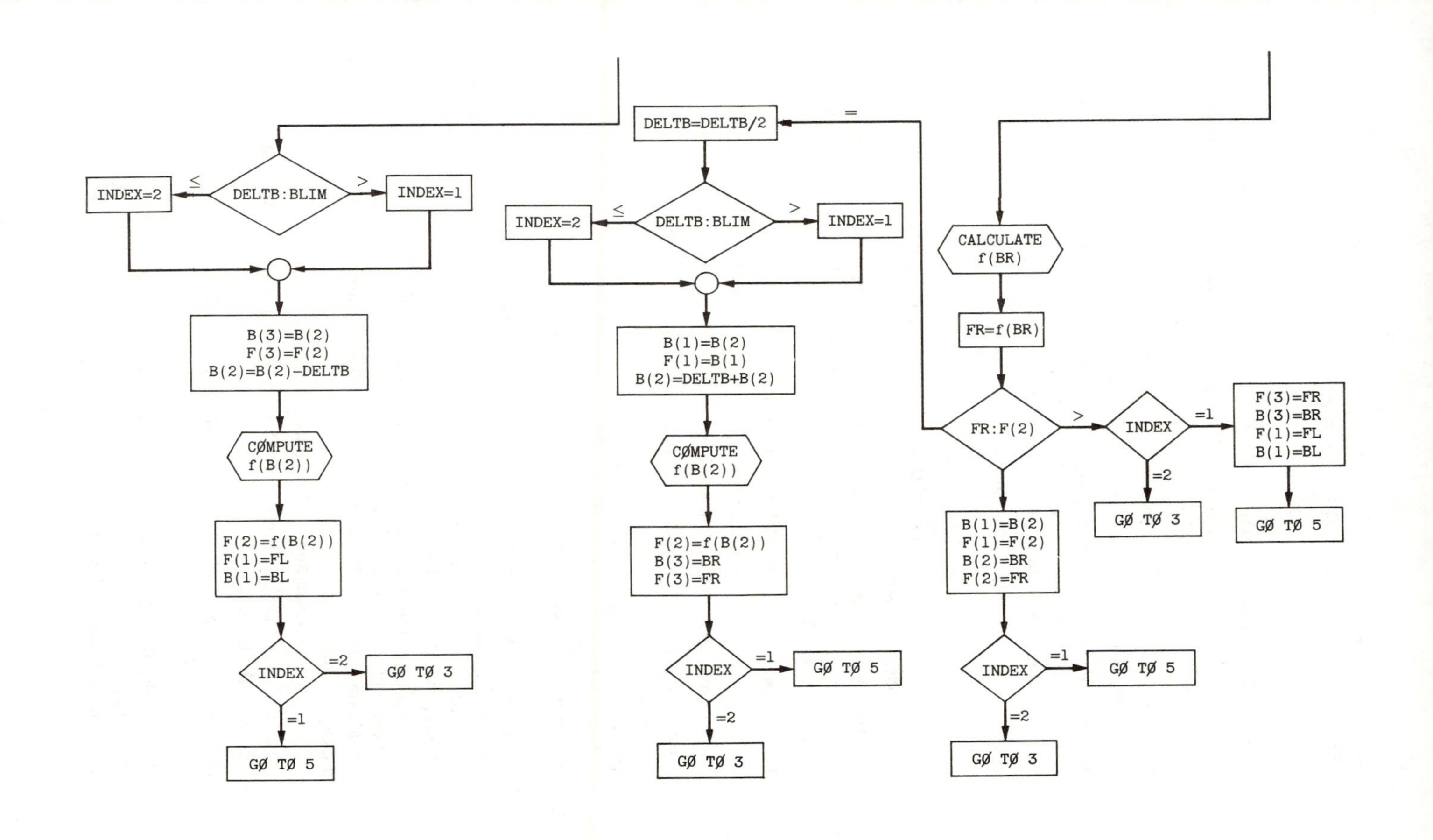
DELTB:BLIM
≤
INDEX=2
>
INDEX=1
B(3)=B(2)
F(3)=F(2)
B(2)=B(2)−DELTB
CØMPUTE
f(B(2))
F(2)=f(B(2))
F(1)=FL
B(1)=BL
INDEX
=2
GØ TØ 3
=1
GØ TØ 5
DELTB=DELTB/2
=
DELTB:BLIM
≤
INDEX=2
>
INDEX=1
B(1)=B(2)
F(1)=B(1)
B(2)=DELTB+B(2)
CØMPUTE
f(B(2))
F(2)=f(B(2))
B(3)=BR
F(3)=FR
INDEX
=1
GØ TØ 5
=2
GØ TØ 3
CALCULATE
f(BR)
FR=f(BR)
FR:F(2)
>
INDEX
=1
F(3)=FR
B(3)=BR
F(1)=FL
B(1)=BL
GØ TØ 5
=2
GØ TØ 3
B(1)=B(2)
F(1)=F(2)
B(2)=BR
F(2)=FR
INDEX
=1
GØ TØ 5
=2
GØ TØ 3

Section 4.2. Interesting discussions of pursuit processes are contained in

BERNHART, A., "Curves of Pursuit II," *Scripta Math.* **XXIII**, 1957, pp. 49–66.

BERNHART, A., "Polygons of Pursuit," *Scripta Math.* **XXIV**, 1958, pp. 23–50.

LITTLEWOOD, J. E., *A Mathematician's Miscellany*, Methuen, London, 1953.

PUCKETTE, C. C., "The Curve of Pursuit," *Math. Gaz.* **37**, 1953, pp. 256–260.

When we allow the rabbit to engage in evasive maneuvers, the problem of formulating what we mean by an optimal pursuit policy becomes quite involved. It can be profitably undertaken using the theory of dynamic programming. See:

BELLMAN, R., *Dynamic Programming*, Princeton University Press, Princeton, N.J., 1957.

Section 4.4. For a discussion of the use of digital computers in the simulation of business operation, see:

BELLMAN, R., C. CLARK, C. CRAFT, D. MALCOLM, and F. RICCIARDI, "On the construction of a multi-person, multistage business game," *Operations Research*, **5**, 1957, pp. 469–503.

For the application of simulation in other connections, see:

QUADE, E. S. (ed.), *Analysis for Military Decisions*, Rand McNally, Chicago, 1965.

For a detailed treatment of the use of finite difference techniques, see:

HENRICI, R., *Discrete Variable Methods in Ordinary Differential Equations*, Wiley, New York, 1962.

The method goes back to Euler and Cauchy.

Section 4.9. A proof of this theorem is provided in Chapter 6.

Section 4.14. We have avoided any analysis of the fundamental concept of stability. An intuitive formulation of the basic problem is the following. Do small changes in the form of the equation or the choice of the system parameters and initial conditions produce small changes in the solution?

Even if we were not interested in this basic question for its intrinsic mathematical content, the question would be forced on us in connection with the use of digital computers with the peculiarly limited kind of arithmetical operations they are capable of.

For analytic formulations of the concept of stability, see the following:

BELLMAN, R., *Stability Theory of Differential Equations*, Dover Publications, New York, 1969.

CODDINGTON, E. S., and N. LEVINSON, *Theory of Ordinary Differential Equations*, McGraw-Hill, New York, 1955.

These books also contain numerous further references on the subject. The book of Henrici, *op. cit.*, contains a thorough discussion of stability as it relates to the numerical solution of differential equations.

Section 4.18. The idea of computation as a control process in which we wish to minimize the overall error is a natural one. Once computation has been conceived of in these terms, the use of feedback and adaptive techniques in general is equally natural.

Section 4.21. For a systematic account of the identification of systems, see:

BELLMAN, R., and R. KALABA, *Quasilinearization and Nonlinear Boundary Value Problems*, American Elsevier, New York, 1965.

Section 4.22. For an account of some search processes encountered in locating the minimum of a function and the point at which the function vanishes, see:

BELLMAN, R., and S. DREYFUS, *Applied Dynamic Programming*, Princeton University Press, Princeton, N.J., 1962.

Further references may be found in this book.

CHAPTER 5

Linear Systems and *N*th-Order Differential Equations

5.1 INTRODUCTION

The great majority of differential equations we meet in practice occur in the form of systems. Instead of a single equation, such as

$$u' = g(u, t), \qquad u(0) = c, \tag{1.1}$$

we encounter a set of equations such as

$$\begin{aligned} \frac{dx_1}{dt} &= g_1(x_1, x_2, \ldots, x_N, t), & x_1(0) &= c_1, \\ \frac{dx_2}{dt} &= g_2(x_1, x_2, \ldots, x_N, t), & x_2(0) &= c_2, \\ &\vdots \\ \frac{dx_N}{dt} &= g_N(x_1, x_2, \ldots, x_N, t), & x_N(0) &= c_N. \end{aligned} \tag{1.2}$$

Fortunately, most of the important methods devised to obtain analytic and computational results for a single equation such as (1.1) carry over to systems of equations. It is also true, however, that the analytic details can often become quite messy unless great care is exerted in the choice of notation. It cannot be sufficiently emphasized that a good notation is a valuable research tool which subtly guides the mind in the desired channels.

The simplest way to explain the appearance of a set of equations such as (1.2) is to point out that physical systems are very seldom described by a single number at any time t. Thus a point particle moving in three-dimensional space is described by means of three position coordinates and three velocity coordinates. A colony of cells may be specified by the size of the total population, but far more precisely and usefully by the number of individual cells with ages between 0 and 1 time units, those between 1 and 2 time units of age, and so on. An ecological system is describable by the sizes of the different plant, insect, and animal populations at any time. An economic system may be specified in terms of the production capacities and stockpiles of different commodities. Examples of this type may be multiplied at will. One of great practical significance is the following.

If we consider an electric circuit such as that in Fig. 5.1, we obtain a coupled pair of linear differential equations for the currents $i_1(t)$ and $i_2(t)$. This is an important point to keep in mind in connection with the use of analog computers based on electric circuits.

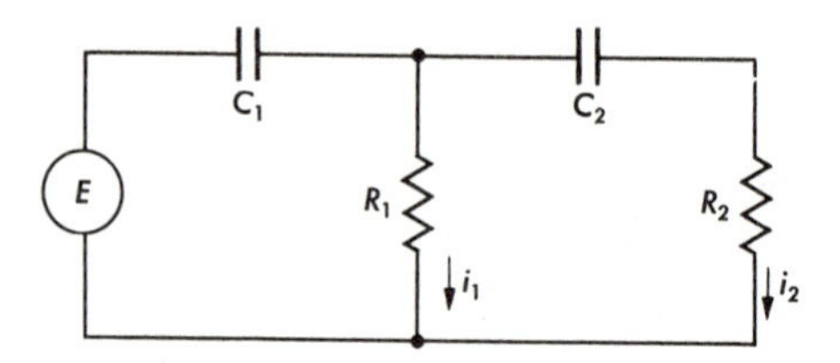

Figure 5.1

As we shall indicate briefly at the end of the chapter, systems of ordinary differential equations arise when we wish to obtain the numerical solution of partial differential equations.

It is natural, as in the one-dimensional case, to consider linear systems first—systems of the form

$$
\begin{aligned}
\frac{dx_1}{dt} &= a_{11}x_1 + a_{12}x_2 + \cdots + a_{1N}x_N, \qquad x_1(0) = c_1, \\
&\vdots \\
\frac{dx_N}{dt} &= a_{N1}x_1 + a_{N2}x_2 + \cdots + a_{NN}x_N, \qquad x_N(0) = c_N,
\end{aligned}
\tag{1.3}
$$

where the coefficients are constants. As we shall see, one simple and direct approach is to convert an equation of this type into an Nth-order linear differential equation of the form

$$
u^{(N)} + a_1 u^{(N-1)} + \cdots + a_N u = 0 \tag{1.4}
$$

for the individual components. We begin with a discussion of what is required to obtain the explicit analytic form of the general solution of Eq. (1.4).

Two methods will be presented. The first, which is a direct extension of the method used for second-order linear differential equations, requires the solution of a system of N simultaneous linear algebraic equations. This has certain disadvantages, regardless of whether we think in terms of hand or computer calculation. Consequently, we shall also present a method based on the use of the Laplace transform, one of the most valuable tools in analysis. This is as good a way as any for the reader to acquaint himself with transform techniques.

Following this, we shall turn to 2×2 systems of linear differential equations. A discussion of general linear systems that has depth or elegance requires vector-matrix theory; we shall therefore avoid such a discussion here. Finally, we shall make some brief remarks concerning power-series techniques and numerical methods based on difference approximations. A detailed treatment of linear systems belongs properly to a second course in differential equations.

5.2 NTH-ORDER LINEAR DIFFERENTIAL EQUATIONS

Consider the Nth-order linear differential equation

$$u^{(N)} + a_1 u^{(N-1)} + \cdots + a_N u = 0, \tag{2.1}$$

where the coefficients are constants, and suppose that we wish to obtain the solution specified by the initial conditions

$$u(0) = c_1,\ u'(0) = c_2,\ \ldots,\ u^{(N-1)}(0) = c_N. \tag{2.2}$$

For the moment, let us set aside questions of existence and uniqueness, which will be treated in Section 5.5, and proceed formally. We begin by looking for particular solutions of Eq. (2.1) of the form $u = e^{rt}$, where r is a constant. Substituting the trial solution in (2.1), we have

$$e^{rt}(r^N + a_1 r^{N-1} + \cdots + a_N) = 0. \tag{2.3}$$

Hence, if r is a root of the polynomial equation

$$r^N + a_1 r^{N-1} + \cdots + a_N = 0, \tag{2.4}$$

then e^{rt} is a solution of (2.1). This polynomial (2.4) is called the *characteristic polynomial*, and the N roots are called the *characteristic roots*. Let us designate them by $r_1, r_2, \ldots, r_N$ and suppose for the moment that they are distinct.

Taking advantage of the linearity of Eq. (2.1), we know that

$$u = b_1 e^{r_1 t} + b_2 e^{r_2 t} + \cdots + b_N e^{r_N t} \tag{2.5}$$

is a solution of (2.1) for any set of constants $b_1, b_2, \ldots, b_N$. It remains for us to choose these constants so as to satisfy the initial conditions (2.2).

Carrying out the differentiations and setting $t = 0$, we obtain the linear system of algebraic equations

$$\begin{aligned} u(0) &= c_1 = b_1 + b_2 + \cdots + b_N, \\ u'(0) &= c_2 = r_1 b_1 + r_2 b_2 + \cdots + r_N b_N, \\ &\vdots \\ u^{(N-1)}(0) &= c_N = r_1^{N-1} b_1 + r_2^{N-1} b_2 + \cdots + r_N^{N-1} b_N. \end{aligned} \tag{2.6}$$

This set of equations possesses a unique solution provided that the determinant

$$\begin{vmatrix} 1 & 1 & \cdots & 1 \\ r_1 & r_2 & \cdots & r_N \\ r_1^{N-1} & r_2^{N-1} & \cdots & r_N^{N-1} \end{vmatrix} = V(r_1, r_2, \ldots, r_N) \tag{2.7}$$

is nonzero.

5.3 THE VANDERMONDE DETERMINANT

Fortunately, the determinant (2.7) is a very famous one, called the *Vandermonde determinant*. It is not difficult to show that it is never zero when the r_i are distinct. Consider first the case $N = 2$. We have

$$\begin{vmatrix} 1 & 1 \\ r_1 & r_2 \end{vmatrix} = r_2 - r_1 \neq 0. \tag{3.1}$$

For $N = 3$, we have

$$\begin{vmatrix} 1 & 1 & 1 \\ r_1 & r_2 & r_3 \\ r_1^2 & r_2^2 & r_3^2 \end{vmatrix} = \begin{vmatrix} 1 & 0 & 0 \\ r_1 & r_2 - r_1 & r_3 - r_1 \\ r_1^2 & r_2^2 - r_1^2 & r_3^2 - r_1^2 \end{vmatrix}, \tag{3.2}$$

where the second determinant is obtained by subtracting the first column in the left-hand determinant from the second and third columns—operations which do not change the value of the determinant.

Factoring out the terms $(r_2 - r_1)$ and $(r_3 - r_1)$, we obtain the expression

$$(3.3)\quad (r_2 - r_1)(r_3 - r_1)\begin{vmatrix} 1 & 0 & 0 \\ r_1 & 1 & 1 \\ r_1^2 & r_2 + r_1 & r_3 + r_1 \end{vmatrix} = (r_2 - r_1)(r_3 - r_1)\begin{vmatrix} 1 & 0 & 0 \\ r_1 & 1 & 1 \\ 0 & r_2 & r_3 \end{vmatrix},$$

where the right-hand determinant results from subtracting r_1 times the second row from the third row of the left-hand determinant. The expression on the right-hand side of Eq. (3.3) is therefore

$$= (r_2 - r_1)(r_3 - r_1)(r_3 - r_2) \neq 0.$$

We can continue inductively in this way to reduce the evaluation of the Vandermonde determinant of order N to the evaluation of a Vandermonde determinant of order $N - 1$. We can thus both obtain the explicit evaluation and show that it is never zero when the r_i are distinct.

EXERCISES

1. Show inductively that

$$V(r_1, r_2, \ldots, r_N) = \prod_{i>j} (r_i - r_j).$$

2. Show that Eqs. (2.6) is equivalent to the statement that

$$k_1c_1 + k_2c_2 + \cdots + k_Nc_N = b_1g(r_1) + b_2g(r_2) + \cdots + b_Ng(r_N),$$

where $g(r) = k_1 + k_2r + \cdots + k_Nr^{N-1}$ is an arbitrary polynomial of degree $N - 1$.

3. Hence determine b_1 by finding the polynomial of degree $N - 1$ with the property that $g(r_1) = 1$, $g(r_i) = 0$, $i \neq 1$.

4. Show that the solution obtained in this way is equivalent to the use of Cramer's rule applied to Eqs. (2.6). [*Hint:* The determinant

$$\begin{vmatrix} 1 & 1 & \cdots & 1 \\ r & r_1 & & r_{N-1} \\ r^2 & r_1^2 & & r_{N-1}^2 \\ \vdots & & & \vdots \\ r^{N-1} & r_1^{N-1} & \cdots & r_{N-1}^{N-1} \end{vmatrix}$$

is a polynomial of degree $N - 1$ which vanishes at the $N - 1$ points $r = r_1, r_2, \ldots, r_{N-1}$.]

5.4 AN ILLUSTRATIVE EXAMPLE

Suppose that we are given the third-order equation

$$u^{(3)} - u^{(2)} - 4u^{(1)} + 4u = 0 \tag{4.1}$$

subject to the initial conditions

$$u(0) = 2, \qquad u'(0) = -1, \qquad u''(0) = 5. \tag{4.2}$$

The differential equation is carefully chosen such that its characteristic equation,

$$r^3 - r^2 - 4r + 4 = 0, \tag{4.3}$$

has the roots $r_1 = 2$, $r_2 = -2$, $r_3 = 1$. Setting

$$u = b_1e^{2t} + b_2e^{-2t} + b_3e^t, \tag{4.4}$$

and using the conditions (4.2), we obtain the linear algebraic equations

$$\begin{aligned} 2 &= b_1 + b_2 + b_3, \\ -1 &= 2b_1 - 2b_2 + b_3, \\ 5 &= 4b_1 + 4b_2 + b_3. \end{aligned} \tag{4.5}$$

This set of equations has the solution $b_1 = 0$, $b_2 = 1$, $b_3 = 1$.

There is no conceptual difficulty in applying the same method to higher-order equations. There is, however, considerable computational labor involved in solving the set of linear algebraic equations corresponding to (4.5) if we are dealing with a fifth- or tenth-order equation. For this reason, in Section 5.6 we shall present an alternative technique based on a rudimentary use of the Laplace transform which is sometimes more convenient.

EXERCISES

1. Let $r^N + a_1r^{N-1} + \cdots + a_N = 0$. Let

$$\begin{aligned} p(D)f &= (D^N + a_1D^{N-1} + \cdots + a_N)f \\ &= f^{(N)} + a_1f^{(N-1)} + \cdots + a_Nf. \end{aligned}$$

Show that

$$p(D)(te^{rt}) = [tp(r) + p'(r)]e^{rt}.$$

[*Hint:* Use Leibniz's rule, $D^k(uv) = (D^ku)v + k(D^{k-1}u)(Dv) + \cdots$]

2. Hence show that if r_1 is a double root of $p(r)$, then $e^{r_1 t}$ and $te^{r_1 t}$ are both solutions of the linear differential equation.
3. Show that if r_1 is a multiple root of order k, then $e^{r_1 t}, te^{r_1 t}, \ldots, t^{k-1}e^{r_1 t}$ are all solutions.
4. Given that r_1 is a multiple root of order k and all the other characteristic roots are simple, show how to solve the initial-value problem

$$u(0) = c_1,\ u'(0) = c_2,\ \ldots,\ u^{(N-1)}(0) = c_N.$$

5.5 EXISTENCE AND UNIQUENESS OF SOLUTION

Now that we have convinced ourselves that we possess a straightforward algorithm for obtaining a solution of the equation

$$u^{(N)} + a_1 u^{(N-1)} + \cdots + a_N u = 0, \tag{5.1}$$

subject to the initial conditions

$$u(0) = c_1,\ u'(0) = c_2,\ \ldots,\ u^{(N-1)}(0) = c_N, \tag{5.2}$$

it remains to prove that this is the only solution. There are many ways of establishing this property of uniqueness which is not very surprising at this stage. Let us present a method which is a straightforward extension of the method used to demonstrate the result for $N = 2$.

The algebraic manipulations can be considerably simplified if Eq. (5.1) is converted into an equivalent system. We write

$$u = u_1,\ u' = u_2,\ \ldots,\ u^{(N-1)} = u_N. \tag{5.3}$$

Then Eq. (5.3), together with (5.1), lead to the system

$$\begin{aligned} u_1' &= u_2, \\ u_2' &= u_3, \\ &\vdots \\ u_N' &= -a_1 u_N - a_2 u_{N-1} - \cdots - a_N u_1, \end{aligned} \tag{5.4}$$

with the initial conditions

$$u_1(0) = c_1,\ u_2(0) = c_2,\ \ldots,\ u_N(0) = c_N. \tag{5.5}$$

The desired uniqueness of the solution of Eqs. (5.4) and (5.5) can therefore be demonstrated if we prove uniqueness for the general linear system

$$
\begin{aligned}
u_1' &= a_{11}u_1 + a_{12}u_2 + \cdots + a_{1N}u_N, \qquad & u_1(0) &= c_1,\\
u_2' &= a_{21}u_1 + a_{22}u_2 + \cdots + a_{2N}u_N, \qquad & u_2(0) &= c_2,\\
&\vdots\\
u_N' &= a_{N1}u_1 + a_{N2}u_2 + \cdots + a_{NN}u_N, \qquad & u_N(0) &= c_N.
\end{aligned} \tag{5.6}
$$

If $(u_1, u_2, \ldots, u_N)$, $(v_1, v_2, \ldots, v_N)$ are two solutions of Eqs. (5.6), then the linearity of the equations and the initial conditions permit us to conclude that the set of w_i,

$$
w_1 = u_1 - v_1,\ w_2 = u_2 - v_2,\ \ldots,\ w_N = u_N - v_N, \tag{5.7}
$$

also satisfies the differential equations (5.6) with the initial conditions

$$
w_1(0) = 0,\ w_2(0) = 0,\ \ldots,\ w_N(0) = 0. \tag{5.8}
$$

To show that this implies that the w_i are identically zero for $t \geq 0$, we use the fundamental device of converting a differential equation into an integral equation. Using (5.6) and the initial conditions of (5.8) we obtain, upon integration, the system of linear integral equations

$$
\begin{aligned}
w_1 &= \int_0^t (a_{11}w_1 + a_{12}w_2 + \cdots + a_{1N}w_N)\,dt_1,\\
&\vdots\\
w_N &= \int_0^t (a_{N1}w_1 + a_{N2}w_2 + \cdots + a_{NN}w_N)\,dt_1.
\end{aligned} \tag{5.9}
$$

From these equations we derive the following relations:

$$
\begin{aligned}
|w_1| &\leq \int_0^t |a_{11}w_1 + a_{12}w_2 + \cdots + a_{1N}w_N|\,dt_1\\
&\leq \int_0^t (|a_{11}|\,|w_1| + |a_{12}|\,|w_2| + \cdots + |a_{1N}|\,|w_N|)\,dt_1,\\
&\vdots\\
|w_N| &\leq \int_0^t (|a_{N1}|\,|w_1| + |a_{N2}|\,|w_2| + \cdots + |a_{NN}|\,|w_N|)\,dt_1.
\end{aligned} \tag{5.10}
$$

Let b denote the largest of the N^2 numbers, $|a_{ij}|$, $i, j = 1, 2, \ldots, N$. Then (5.10) leads to the relations

$$
\begin{aligned}
|w_1| &\leq b \int_0^t (|w_1| + |w_2| + \cdots + |w_N|)\,dt_1,\\
&\vdots\\
|w_N| &\leq b \int_0^t (|w_1| + |w_2| + \cdots + |w_N|)\,dt_1.
\end{aligned} \tag{5.11}
$$

Adding these inequalities, we have

$$(5.12)\qquad (|w_1| + |w_2| + \cdots + |w_N|) \leq Nb \int_0^t (|w_1| + \cdots + |w_N|)\, dt_1.$$

We have already seen in Chapter 2 that, for $t \geq 0$, this inequality implies that

$$(5.13)\qquad |w_1| + |w_2| + \cdots + |w_N| = 0.$$

Equation (5.13) in turn implies that each w_i is zero for $t \geq 0$. We have thus proved the uniqueness theorem.

EXERCISES

1. Show that the same method can be used to establish the uniqueness of solution for the system of differential equations

$$u_i' = \sum_{j=1}^{N} a_{ij}(t)u_j, \qquad u_i(0) = c_i, \qquad i = 1, 2, \ldots, N,$$

under the assumption that the coefficient functions $a_{ij}(t)$ are continuous for $t \geq 0$. The existence of a solution will be demonstrated in Chapter 6. [*Hint:* Replace the number b in (5.11) by the largest of the numbers m_{ij}, $i, j = 1, 2, \ldots, N$, where m_{ij} is the maximum of the functions $|a_{ij}(t)|$ over a fixed interval $[0, t_0]$.]

2. What is wrong with the following argument? "Consider the one-dimensional equation $u_1' = a_1u_1$, $u_1(0) = 0$, where a_1 is a constant. Then $u_1'(0) = 0$. Differentiating both sides of the equation, we have $u_1'' = a_1u_1'$, which means that $u_1''(0) = 0$. Continuing in this fashion, we see that all derivatives of u_1 are zero at $t = 0$. Hence, $u_1 = 0$ for $t \geq 0$."

5.6 THE LAPLACE TRANSFORM

Consider the function $F(s)$ defined by the relation

$$(6.1)\qquad F(s) = \int_0^\infty e^{-st}u(t)\, dt,$$

where $u(t)$ is a solution of a linear differential equation with constant coefficients. Since we know that $u(t)$ has the form

$$b_1e^{\lambda_1 t} + b_2e^{\lambda_2 t} + \cdots + b_Ne^{\lambda_N t},$$

where in the limiting case the coefficients can be polynomials in t, we can ensure convergence of the integral by taking s sufficiently large.

If $u(t)$ is an exponential of the form e^{bt}, then $F(s)$ has a very simple form,

$$\int_0^\infty e^{-st}e^{bt}\,dt = \frac{1}{s-b}, \tag{6.2}$$

provided that s is sufficiently large. Conversely, if we are told that $1/(s - b)$ is the Laplace transform of an exponential function, then we know that this function must be e^{bt}.

If $u(t) = te^{bt}$, then

$$F(s) = \int_0^\infty e^{-st}te^{bt}\,dt = \frac{1}{(s-b)^2}; \tag{6.3}$$

and, generally, a simple integration by parts shows that

$$\int_0^\infty e^{-st}t^{k-1}e^{bt}\,dt = \frac{(k-1)!}{(s-b)^k}, \tag{6.4}$$

for $k = 1, 2, \ldots$ (Let $0! = 1$, as is customary.) Thus, if we encounter a term such as $1/(s - b)^k$ we associate it with the term $t^{k-1}e^{bt}/(k - 1)!$.

The derivative of the Laplace transform possesses a remarkable and most important property. Consider the integral

$$\int_0^\infty e^{-st}u'(t)\,dt, \tag{6.5}$$

and let us proceed to integrate by parts. We have

$$\int_0^\infty e^{-st}u'(t)\,dt = [e^{-st}u(t)]_0^\infty + s\int_0^\infty e^{-st}u(t)\,dt. \tag{6.6}$$

If s is large enough so that the integral (6.1) converges, then the contribution at $t = \infty$ drops out, and we have the important formula

$$\int_0^\infty e^{-st}u'(t)\,dt = -u(0) + s\int_0^\infty e^{-st}u(t)\,dt. \tag{6.7}$$

This relation enables us to solve the equation

$$u'(t) - bu(t) = 0, \qquad u(0) = c, \tag{6.8}$$

in the following simple fashion. From Eq. (6.8) we have

$$\begin{gathered}\int_0^\infty e^{-st}[u'(t) - bu(t)]\,dt = 0,\\ \int_0^\infty e^{-st}u'(t)\,dt - b\int_0^\infty e^{-st}u(t)\,dt = 0.\end{gathered} \tag{6.9}$$

Using Eq. (6.7), we find that (6.9) becomes

$$s \int_0^\infty e^{-st}u(t)\,dt - c - b \int_0^\infty e^{-st}u(t)\,dt = 0. \tag{6.10}$$

Hence

$$\int_0^\infty e^{-st}u(t)\,dt = \frac{c}{s-b}. \tag{6.11}$$

At this point, we face the familiar bugaboo of uniqueness of solution. There may conceivably be many functions $u(t)$ which satisfy the integral equation (6.11). How do we know which one to pick and, in any case, how to determine it? The answer in this case is simple. We are not interested in obtaining the general solution of (6.11). We wish to use the relation solely to match constants. We already know that the general solution of (6.8) is an exponential, $ae^{\lambda t}$. Since

$$\int_0^\infty ae^{\lambda t}e^{-st}\,dt = \frac{a}{s-\lambda},$$

we see that $a = c, \lambda = b$. This is a most important point as far as our application of the Laplace transform is concerned, since it enables us to bypass some thorny questions of analysis.

Emboldened by this success, suppose that we wish to solve the second-order linear differential equation

$$u'' + 3u' + 2u = 0, \qquad u(0) = 2, \qquad u'(0) = -3. \tag{6.12}$$

We begin by repeating the trick used in (6.7). Repeated integration by parts yields

$$\begin{aligned} \int_0^\infty e^{-st}u''\,dt &= e^{-st}u'\Big]_0^\infty + s\int_0^\infty e^{-st}u'\,dt \\ &= -u'(0) + s\left[-u(0) + s\int_0^\infty e^{-st}u\,dt\right] \\ &= -u'(0) - su(0) + s^2\int_0^\infty e^{-st}u(t)\,dt. \end{aligned} \tag{6.13}$$

Hence, from (6.12) we have

$$\begin{aligned} 0 &= \int_0^\infty e^{-st}(u'' + 3u' + 2u)\,dt \\ &= \int_0^\infty e^{-st}u''\,dt + 3\int_0^\infty e^{-st}u'\,dt + 2\int_0^\infty e^{-st}u\,dt \\ &= 3 - 2s + s^2\int_0^\infty e^{-st}u\,dt \\ &\quad + 3\left[-2 + s\int_0^\infty e^{-st}u(t)\,dt\right] + 2\int_0^\infty e^{-st}u\,dt. \end{aligned} \tag{6.14}$$

Collecting terms, we obtain the relation

$$\int_0^\infty e^{-st}u(t)\,dt = \frac{2s+3}{s^2+3s+2}. \tag{6.15}$$

How do we recover the solution $u(t)$ from (6.15)?

5.7 PARTIAL FRACTIONS

The answer again is simple. We invoke the important device of the resolution of a rational function into partial fractions. The factors of the polynomial in the denominator are $(s + 1)$ and $(s + 2)$. Hence we write

$$\frac{2s+3}{s^2+3s+2} = \frac{a_1}{s+2} + \frac{a_2}{s+1}, \tag{7.1}$$

where a_1 and a_2 are constants to be found. There are several ways of performing this decomposition. One way is the following. To determine a_1, we multiply Eq. (7.1) through by $s + 2$ and then let $s = -2$. We have

$$\frac{2s+3}{s+1} = a_1 + a_2\frac{s+2}{s+1}. \tag{7.2}$$

Hence

$$a_1 = 1. \tag{7.3}$$

Similarly, we see that $a_2 = 1$. Returning to (6.15), we see that

$$\int_0^\infty e^{-st}u(t)\,dt = \frac{1}{s+2} + \frac{1}{s+1}. \tag{7.4}$$

Since we are looking for a solution of the form

$$u = b_1e^{\lambda_1 t} + b_2e^{\lambda_2 t},$$

a matching of parameters yields

$$u = e^{-2t} + e^{-t}. \tag{7.5}$$

The method of partial fractions encounters some slight difficulties when multiple roots are present, as the reader may recall from integral calculus. If, for example, we encounter the function

$$F(s) = \frac{1}{s^3+s^2-s-1}$$

with the factorization $(s^3 + s^2 - s - 1) = (s + 1)^2(s - 1)$, we write

$$\frac{1}{s^3 + s^2 - s - 1} = \frac{a_1}{(s + 1)^2} + \frac{a_2}{s + 1} + \frac{a_3}{s - 1}. \tag{7.6}$$

Using the foregoing technique we readily obtain

$$a_3 = \tfrac{1}{4}, \qquad a_1 = -\tfrac{1}{2}. \tag{7.7}$$

What about a_2? To obtain this missing value, we write

$$\frac{1}{s^3 + s^2 - s - 1} = \frac{-\frac{1}{2}}{(s + 1)^2} + \frac{a_2}{s + 1} + \frac{\frac{1}{4}}{s - 1}. \tag{7.8}$$

Now choose a convenient value of s, say $s = 0$, to determine a_2. We have

$$\frac{1}{-1} = -\frac{1}{2} + a_2 - \frac{1}{4}, \tag{7.9}$$

whence $a_2 = -\frac{1}{4}$. Thus, finally, we get

$$\frac{1}{s^3 + s^2 - s - 1} = \frac{-\frac{1}{2}}{(s + 1)^2} - \frac{\frac{1}{4}}{s + 1} + \frac{\frac{1}{4}}{s - 1}. \tag{7.10}$$

It follows that the corresponding function in the t-domain is given by

$$u(t) = -\frac{te^{-t}}{2} - \frac{e^{-t}}{4} + \frac{e^{t}}{4}. \tag{7.11}$$

EXERCISES

1. Use the Laplace transform to solve the following equations:

   a) $u' + u = t,\ u(0) = 1$
   b) $u' + 2u = \cos 3t,\ u(0) = 1$
   c) $u'' + u = t,\ u(0) = 0,\ u'(0) = 1$
   d) $u'' + u = \sin \omega t,\ u(0) = 0,\ u'(0) = 1$. Distinguish between the cases $\omega \neq 1$, $\omega = 1$.
   e) $u^{(4)} + 5u^{(2)} + u = 0,\ u(0) = u'(0) = 1,\ u^{(2)}(0) = u^{(3)}(0) = 0.$

2. Obtain the coefficients a_1 and a_2 in Eq. (7.1) in the following fashion. From

$$\frac{2s + 3}{s^2 + 3s + 2} = \frac{a_1}{s + 2} + \frac{a_2}{s + 1}$$

we have

$$\frac{2s+3}{s^2+3s+2} = \frac{a_1(s+1)+a_2(s+2)}{(s+2)(s+1)}.$$

Since the numerator on the right-hand side must be identically equal to the numerator on the left-hand side, we must have $a_1 + a_2 = 2, a_1 + 2a_2 = 3$.

3. Similarly, show that the decomposition

$$\frac{1}{s^3+s^2-s-1} = \frac{a_1}{(s+1)^2} + \frac{a_2}{s+1} + \frac{a_3}{s-1}$$

leads to the equations

$$a_2 + a_3 = 0, a_1 + 2a_3 = 0, -a_1 - a_2 + a_3 = 1,$$

and thus determine a_1, a_2, a_3.

5.8 DISCUSSION

The advantages of the Laplace transform in this instance reside in the fact that it enables us to avoid the numerical solution of a system of linear algebraic equations such as Eqs. (2.6). Let the ambitious reader tackle merely a 6×6 system and he will thereafter appreciate the point. Regardless of which method we employ, in order to obtain an explicit analytic solution we must determine the characteristic roots. Neither this task nor that of solving a linear system of algebraic equations need be faced if we content ourselves with a purely numerical solution using the techniques described in Chapter 4, or if we employ a numerical technique for finding $u(t)$ given its Laplace transform.

At all accounts, let us emphasize the fact that arithmetic is truly difficult and that it is not beneath the dignity of even the finest mathematician to devote some of his talent and ingenuity to circumventing it.

5.9 LINEAR SYSTEMS

Consider the system of linear differential equations

$$\begin{aligned} x_1' &= 2x_1 + 3x_2, \qquad x_1(0) = 1, \\ x_2' &= x_1 + x_2, \qquad x_2(0) = 2. \end{aligned} \tag{9.1}$$

A simple and direct method of solution is one which uses the technique of elimination. To eliminate x_2, for instance, we write (9.1) in the form

$$\left(\frac{d}{dt}-2\right)x_1 = 3x_2, \qquad x_1(0) = 1,$$

(9.2)

$$\left(\frac{d}{dt}-1\right)x_2 = x_1, \qquad x_2(0) = 2.$$

Then we have

$$\left(\frac{d}{dt}-1\right)\left(\frac{d}{dt}-2\right)x_1 = 3\left(\frac{d}{dt}-1\right)x_2 = 3x_1, \tag{9.3}$$

or

$$x_1'' - 3x_1' - x_1 = 0. \tag{9.4}$$

The initial conditions are

$$x_1(0) = 1, \qquad x_1'(0) = 2(1) + 3(2) = 8. \tag{9.5}$$

Having obtained $x_1(t)$, we can obtain x_2 using the second equation in (9.1), or we can eliminate x_1 and obtain equations corresponding to (9.4) and (9.5) for x_2.

EXERCISES

1. Solve the system

$$x_1' = x_1 + x_2 + 2x_3, \qquad x_2' = x_1 + 2x_2 + x_3,$$
$$x_3' = 2x_1 + x_2 + x_3,$$

where

$$x_1(0) = x_2(0) = x_3(0) = 1.$$

2. Determine x_1 and x_2 for

$$x_1' = 2x_1 + 3x_2 + \sin \omega t, \qquad x_2' = x_1 + x_2,$$

where

$$x_1(0) = x_2(0) = 0.$$

3. Use the Laplace transform to obtain the solutions of the foregoing equations.

5.10 AN ALTERNATIVE APPROACH

Suppose we proceed in the following fashion. Starting with

$$x_1' = 2x_1 + 3x_2, \qquad x_2' = x_1 + x_2, \tag{10.1}$$

let us attempt to find a particular solution of the form $x_1 = e^{rt}c_1$, $x_2 = e^{rt}c_2$, where r, c_1, c_2 are constants. Upon substituting in (10.1), we have

$$rc_1 = 2c_1 + 3c_2, \qquad rc_2 = c_1 + c_2, \tag{10.2}$$

or

$$(2 - r)c_1 + 3c_2 = 0, \qquad c_1 + (1 - r)c_2 = 0. \tag{10.3}$$

A necessary and sufficient condition that this system of linear homogeneous equations have a nontrivial solution, i.e., that c_1, c_2 be not both zero, is given by the determinantal equation

$$\begin{vmatrix} 2 - r & 3 \\ 1 & 1 - r \end{vmatrix} = 0. \tag{10.4}$$

The two roots of this quadratic equation are distinct. Corresponding to each root we can obtain values for the constants c_1 and c_2 in (10.2) and then, by superposition, the general solution of (10.1). We shall not pursue this approach any further since any clear exposition of the application of this important method to higher-order systems requires the use of matrix theory. References will be given at the end of this chapter for the reader who desires to study this approach in detail.

5.11 POWER-SERIES SOLUTIONS

There is no difficulty in extending the techniques presented in Chapter 3 and applying them to higher-order linear differential equations and systems of differential equations. As we pointed out in Section 5.10, the use of vector-matrix notation is essential in discussing higher-order equations and systems of equations. Consequently, we urge the reader to devote a few months to mastering the fundamentals of matrix theory—which is a basic tool in all advanced mathematics—and then to return to the fray.

5.12 SOME REMARKS CONCERNING COMPUTATIONAL TECHNIQUES

The basic ideas involved in the numerical solution of systems of ordinary differential equations by means of digital computers are the same as those discussed in Chapter 4 in connection with first-order equations. Naturally, the successful implementation of these ideas requires a certain amount of training and experience.

The important point for the reader to remember is that standard methods are now available to handle systems of quite large dimensions, 100, 500, or 1000. Consequently, at the present time, we can use quite realistic mathematical models of physical processes with confidence that numerical answers can be obtained.

Each five years over the past twenty has seen a remarkable increase in both the versatility and capability of the computer. We can expect a similar rate of progress to be sustained over the next twenty years. Thus by 1970 we can confidently expect to be able to solve with ease, speed, and accuracy systems involving ten thousand simultaneous nonlinear ordinary differential equations subject to initial conditions. By 1980, this figure should have grown to one million.

This ability to do arithmetic dramatically changes the field of applied mathematics and opens unimaginable vistas in the domain of science.

5.13 PARTIAL DIFFERENTIAL EQUATIONS

By virtue of what was said in the foregoing section, it is feasible to reduce the numerical solution of many important classes of partial differential equations to the problem of solving large systems of ordinary differential equations. Consider, for example, the nonlinear heat equation

$$\frac{\partial u}{\partial t} = \frac{\partial^2 u}{\partial x^2} + u^2, \qquad 0 < x < 1, \quad t > 0, \tag{13.1}$$

where

$$\begin{aligned} u(x, 0) &= g(x), \\ u(0, t) &= u(1, t) = 0, \qquad t \geq 0. \end{aligned} \tag{13.2}$$

Let us divide the x-interval into N parts and use the approximation

$$\frac{\partial^2 u}{\partial x^2} \cong \frac{u(x + \Delta, t) + u(x - \Delta, t) - 2u(x, t)}{\Delta^2}, \tag{13.3}$$

where x assumes the values $\Delta, 2\Delta, \ldots, (N-1)\Delta$. Then Eq. (13.1) takes the form

$$\frac{\partial u(x,t)}{\partial t} = \frac{u(x+\Delta,t) + u(x-\Delta,t) - 2u(x,t)}{\Delta^2} + u(x,t)^2. \tag{13.4}$$

If we write

$$u(k\Delta, t) = u_k(t), \tag{13.5}$$

then (13.4) becomes

$$u_k'(t) = \frac{u_{k+1} + u_{k-1} - 2u_k}{\Delta^2} + u_k^2, \tag{13.6}$$

for $k = 1, 2, \ldots, N-1$, a system of $N-1$ ordinary differential equations. From the conditions (13.2) we have

$$u_0(t) = u_N(t) = 0. \tag{13.7}$$

The initial conditions are

$$u_k(0) = g(k\Delta). \tag{13.8}$$

The choice of $\Delta = 1/N$ can be made adaptive in the sense that we can let the computer solve the system of (13.6) for different values of N, N_0, $2N_0$, $4N_0, \ldots$, until we obtain a desired accuracy in the solution.

This, of course, is only the beginning of the story. There are many interesting and difficult analytic questions to answer before we can employ methods of this type with confidence, as many a computer-user has found to his sorrow. It is nevertheless true that the existence of electronic computers has made it possible to attack problems of this kind with high hopes of success.

MISCELLANEOUS EXERCISES

---

1. Consider the function

$$w(t) = \int_0^t u(t_1)v(t-t_1)\,dt_1.$$

Show that under reasonable assumptions concerning u and v we have $L(w) = L(u)L(v)$, where $L(u)$ is the Laplace transform of u. [*Hint:* Consider

$$\int_0^\infty e^{-st}\left[\int_0^t u(t_1)v(t-t_1)\,dt_1\right] dt$$

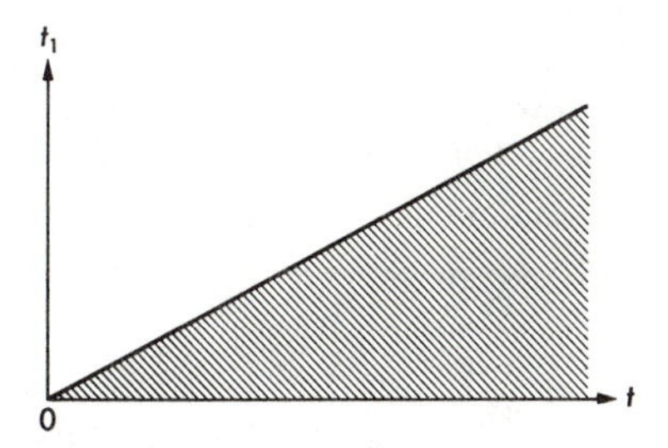

Figure 5.2

as a double integral over the shaded region shown in Fig. 5.2, and invert the order of integration.]

2. Let u satisfy the equation $u' + au = f(t)$, $u(0) = c$. Show that

$$L(u) = \frac{L(f)}{s+a} + \frac{c}{s+a}.$$

Hence, using the result of Exercise 1, show that

$$u = ce^{-at} + \int_0^t e^{-a(t-t_1)} f(t_1)\, dt_1.$$

3. Similarly obtain the solution of

$$u'' + a_1 u' + a_2 u = f(t), \qquad u(0) = c_1, \qquad u'(0) = c_2.$$

4. Show that, generally, the solution of

$$u^{(n)} + a_1 u^{(n-1)} + \cdots + a_n u = f(t),$$

where $u(0) = u'(0) = \cdots = u^{(n-1)}(0) = 0$, has the form

$$u(t) = \int_0^t k(t - t_1) f(t_1)\, dt_1,$$

and determine the form of $k(t)$.

5. Without worrying about existence, uniqueness, and the legitimacy of the operations, show that if $u(x, t)$ satisfies the linear partial differential equation

$$u_t = u_{xx}, \qquad t > 0, \qquad 0 < x < 1,$$
$$u(x, 0) = g(x), \qquad u(0, t) = u(1, t) = 0,$$

then $v(x, s) = L(u)$ satisfies the ordinary differential equation

$$v'' - sv = -g(x), \qquad v(0, s) = v(1, s) = 0,$$

where $' = \partial/\partial x$. Hence determine v.

6. Consider the first-order differential equation $du/dt = au - b$, $u(0) = c$. Show that if $u(t)$ has a limit as $t \to \infty$, then this limit is b/a.
7. Show that $u(t)$ has a limit as $t \to \infty$ provided that $a < 0$.
8. Consider the linear system

$$x_1' = a_1x_1 + a_2x_2 - a_3, \qquad x_2' = b_1x_1 + b_2x_2 - b_3,$$

where $x_1(0) = c_1$, $x_2(0) = c_2$. Show that if $x_1(t)$, $x_2(t)$ possess limits as $t \to \infty$, then these limits are the solution of the linear algebraic system

$$a_1x_1 + a_2x_2 - a_3 = 0, \qquad b_1x_1 + b_2x_2 - b_3 = 0.$$

9. Show that the limits referred to in Exercise 8 exist if the roots of the determinantal equation

$$\begin{vmatrix} a_1 - \lambda & a_2 \\ b_1 & b_2 - \lambda \end{vmatrix} = 0$$

have negative real parts.
10. Consider the differential equation

$$\frac{du}{dt} = g(u), \qquad u(0) = c.$$

Show that if $u(t)$ possesses a limit as $t \to \infty$, then this limit is a root of the equation $g(u) = 0$.
11. Consider the simultaneous differential equations

$$\frac{dx_1}{dt} = g(x_1, x_2), \qquad \frac{dx_2}{dt} = h(x_1, x_2).$$

Show that if x_1 and x_2 possess limits as $t \to \infty$, then these limits are solutions of the simultaneous equations $g(x_1, x_2) = 0$, $h(x_1, x_2) = 0$. (The idea of regarding the solutions of a system of equations as the limiting values of the solutions of a system of differential equations is a very powerful one. It is the basis of the gradient technique.)
12. Let u_1 and u_2 be distinct solutions of the differential equation

$$u'' + a_1(t)u' + a_2(t)u = 0.$$

Then the general solution of

$$u''' + 3a_1u'' + (2a_1^2 + a_1' + 4a_2)u' + (4a_1a_2 + 2a_2')u = 0$$

is given by $u = a_1u_1^2 + c_2u_1u_2 + c_3u_2^2$. (This result is due to Appell.)

13. The determinant

$$W(t) = \begin{vmatrix} u_1 & u_2 \\ u_1' & u_2' \end{vmatrix}$$

is called the Wronskian of the functions u_1 and u_2. If u_1 and u_2 are solutions of $u'' + a_1(t)u' + a_2(t)u = 0$, show that

$$dW/dt = -a_1(t)W.$$

[*Hint:* Recall the rule for differentiating a determinant.]

14. Hence show that

$$W(t) = \exp\left[-\int_0^t a_1(t_1)\, dt_1\right] W(0).$$

Deduce that if $a_1(t)$ is continuous for $t \geq 0$, then $W(t)$ is zero at a point $t > 0$ only if it is identically zero.

15. If u_1 and u_2 are principal solutions of

$$u'' + a_1(t)u' + a_2(t)u = 0,$$

then the equations $x_1u_1(t) + x_2u_2(t) = 0$, $x_1u_1'(t) + x_2u_2'(t) = 0$ possess the unique solution $x_1 = x_2 = 0$.

16. Using the result of Exercise 14 and the equation $u'' + u = 0$, deduce that

$$\sin^2 t + \cos^2 t = 1.$$

17. Obtain the analog of the results obtained in Exercises 13 and 14 for the nth-order linear differential equation

$$u^{(n)} + a_1(t)u^{(n-1)} + \cdots + a_n(t)u = 0.$$

18. Show that

$$\begin{vmatrix} u'' & u' & u \\ u_1'' & u_1' & u_1 \\ u_2'' & u_2' & u_2 \end{vmatrix} = 0$$

is a second-order linear differential equation with the solutions $u = u_1$ and $u = u_2$.

19. Using the relation given in Exercise 18, obtain relations between the solutions u_1, u_2 and the coefficients a_1 and a_2 of the equation $u'' + a_1u' + a_2u = 0$ analogous to those connecting the roots and coefficients of an algebraic equation.

20. Show that

$$\frac{d^2u}{dt^2} + a_1(t)\frac{du}{dt} + a_2(t)u = \frac{W}{u_1}\frac{d}{dt}\left[\frac{u_1^2}{W}\frac{d}{dt_1}\left(\frac{u}{u_1}\right)\right]$$

if u_1 and u_2 are two solutions for which $W(0) \neq 0$. (This result and its generalization to nth order equations are treated in G. Polya and G. Szego, *Aufgaben und Lehrsätze aus der Analysis*, Dover, New York, 1945, Vol II, p. 113.)

BIBLIOGRAPHY AND COMMENTS

Section 5.1. For an introduction to matrix theory with applications to the theory of linear differential equations, see:

BELLMAN, R., *Introduction to Matrix Analysis*, McGraw-Hill, New York, 1960.

Section 5.2. For an introductory account of the Laplace transform, see:

BELLMAN, R., R. KALABA, and J. LOCKETT, *Numerical Inversion of the Laplace Transform with Applications to Biology, Economics, Engineering, and Physics*, American Elsevier, New York, 1966.

Section 5.10. See the book on matrix theory cited above.

CHAPTER 6

Existence and Uniqueness Theorems

6.1 INTRODUCTION

In the previous chapters, we focused our attention on a presentation of some of the basic analytic and computational techniques of the theory of differential equations, striving resolutely to keep the rigorous details to a minimum. It was not that we felt that rigor was not important, or that the existence of computers obviated the need for careful analysis. Rather, it was a desire to make the basic principles as clear as possible and to motivate the rigorous treatment.

In this concluding chapter, we wish to fill in some noteworthy gaps in the material presented so far. First, we wish to indicate how the uniqueness of solution is established for general classes of nonlinear equations under suitable, simple hypotheses using a straightforward extension of the method developed to treat linear equations. Then we wish to show how to employ that versatile and powerful tool of analysis—the method of successive approximations—to demonstrate the existence of a solution. As both an analytical and computational workhorse, this method occupies a unique position. The method of quasilinearization, yielding a sequence of successive approximations which converges quite rapidly, is briefly discussed.

Finally, we shall sketch a proof of the fact that the finite difference method used in Chapter 4 to obtain numerical results is indeed valid.

6.2 UNIQUENESS

Let us introduce the technique we will employ by considering first the specific equation

$$u' = u^2 + t, \qquad u(0) = 1. \tag{2.1}$$

Our aim is to show that there is at most one solution in any t-interval $0 \leq t \leq a$. By a solution we mean a function $u(t)$ which is defined over this closed interval and, in addition, possesses a derivative $u'(t)$ at every point of this interval which satisfies the equation (2.1). In general, if the equation is nonlinear, then a is not arbitrarily large. (In Section 6.6 we will show how an estimate for a can be obtained. Let us suppose that there are two such solutions, u and v, and then obtain a contradiction. Thus we have

$$v' = v^2 + t, \qquad v(0) = 1. \tag{2.2}$$

Subtracting Eq. (2.2) from Eq. (2.1), we have

$$(u - v)' = u^2 - v^2. \tag{2.3}$$

Integrating between 0 and t, we obtain

$$u - v = \int_0^t (u^2 - v^2)\, dt_1, \tag{2.4}$$

since $u - v = 0$ at $t = 0$. Hence

$$\begin{aligned} |u - v| &= \left| \int_0^t (u^2 - v^2)\, dt_1 \right| \leq \int_0^t |u^2 - v^2|\, dt_1 \\ &= \int_0^t |u - v|\, |u + v|\, dt_1, \end{aligned} \tag{2.5}$$

by the identity

$$u^2 - v^2 = (u - v)(u + v).$$

We are tacitly assuming that u and v are real, but as we shall see, the proof goes through without any change of detail if $|\cdot|$ denotes the absolute value of a complex number. Since u and v are continuous in the closed interval $[0, a]$, they are bounded there. Let b denote the maximum value of $|u|$, $|v|$ for t in $[0, a]$. Then

$$|u + v| \leq |u| + |v| \leq 2b. \tag{2.6}$$

Thus (2.5) yields

$$|u - v| \leq 2b \int_0^t |u - v|\, dt_1. \tag{2.7}$$

From this relation we wish to conclude that $u - v \equiv 0$. There are many ways of doing this, as we indicated in the foregoing chapters. For the sake of completeness let us present a proof here. A particularly elegant approach is based on the following lemma.

Lemma. *If $f(t)$, $g(t)$ are nonnegative for $t \geq 0$ and $c > 0$, then*

$$f(t) \leq c + \int_0^t f(t_1)g(t_1)\,dt_1 \tag{2.8}$$

implies that

$$f(t) \leq c \exp\left[\int_0^t g(t_1)\,dt_1\right]\cdot \tag{2.9}$$

For the proof of this lemma, see Exercise 12 of Section 2.4.

Returning to the relation (2.7), we observe that it implies

$$|u - v| \leq \epsilon + 2b\int_0^t |u - v|\,dt_1 \tag{2.10}$$

for any $\epsilon > 0$. Applying the preceding lemma, we have

$$|u - v| \leq \epsilon \exp\left[2b\int_0^t dt_1\right] = \epsilon e^{2bt}. \tag{2.11}$$

Since this relation holds for any $\epsilon > 0$, we see that $|u - v| \equiv 0$, and therefore $u = v$, which establishes the uniqueness of the solution for (2.1). It is interesting to note that we proved uniqueness of solution of the differential equation by transforming it into an integral equation. This is a standard device in the theory of differential equations as previously noted.

6.3 THE EQUATION $u' = g(u, t)$, $u(0) = c$

Let us now see what is required to establish the corresponding result for the general first-order equation

$$u' = g(u, t), \qquad u(0) = c. \tag{3.1}$$

Let u and v be two solutions. Then, as in the last section, we have

$$(u - v) = \int_0^t [g(u, t_1) - g(v, t_1)]\,dt_1, \tag{3.2}$$

and thus

$$|u - v| \leq \int_0^t |g(u, t_1) - g(v, t_1)|\,dt_1. \tag{3.3}$$

What we now wish to do is to bound the function $|g(u, t) - g(v, t)|$ by a term involving $|u - v|$. We accomplished this in Section 6.2 by the simple expedient of factoring $u^2 - v^2$. For the general case, we need a mean-value theorem,

$$g(u, t) - g(v, t) = (u - v)g_u(\theta, t), \tag{3.4}$$

where θ lies between u and v, and g_u denotes the partial derivative with respect to u. In the equation (3.4) we are implicitly assuming that u and v are real. If they are complex, we use the technique of Section 6.5. If we suppose that $|g_u|$ is bounded for $0 \le t \le t_0$ and $|u| \le b$, then the argument proceeds as before. We have

$$|g(u, t) - g(v, t)| = |u - v|\,|g_u(\theta, t)| \le b_1|u - v|, \tag{3.5}$$

where b_1 is the bound on g_u. Hence the relation (3.3) yields

$$|u - v| \le b_1 \int_0^t |u - v|\, dt_1. \tag{3.6}$$

Following the steps given in Section 6.2, we see that $u \equiv v$.

If an inequality such as (3.5) holds in a region $0 \le t \le t_0$, $|u|, |v| \le b$, then the function $g(u, t)$ is said to satisfy a *Lipschitz condition*. The simplest sufficient condition for this condition to hold is that $g(u, t)$ possess a partial derivative with respect to u for $0 \le t \le t_0$, $|u| \le b$. This is clearly the case if $g(u, t)$ is a polynomial in u and t. Hence we see that we have established the theorem given in Section 4.9.

6.4 HIGHER-ORDER EQUATIONS

To obtain existence and uniqueness theorems for a higher-order differential equation, such as

$$u'' = g(u', u, t), \tag{4.1}$$

it is simplest to convert the equation into the system

$$u' = v, \qquad v' = g(v, u, t), \tag{4.2}$$

and then employ a general uniqueness theorem concerning systems. Let us briefly sketch the procedure of obtaining this immediate generalization of the result of Section 6.3.

6.5 SYSTEMS

Consider the system of differential equations

$$
(5.1)\qquad
\begin{aligned}
\frac{dx_1}{dt} &= g_1(x_1, x_2, \ldots, x_N), \qquad x_1(0) = c_1,\\
&\ \vdots\\
\frac{dx_N}{dt} &= g_N(x_1, x_2, \ldots, x_N), \qquad x_N(0) = c_N.
\end{aligned}
$$

Let $(x_1, x_2, \ldots, x_N)$, $(y_1, y_2, \ldots, y_N)$ be two solutions of (5.1) in the interval $0 \le t \le t_0$. Then, as before, subtracting the corresponding equations and integrating, we have

$$
(5.2)\qquad
\begin{aligned}
x_1 - y_1 &= \int_0^t [g_1(x_1, x_2, \ldots, x_N) - g_1(y_1, y_2, \ldots, y_N)]\, dt_1,\\
&\ \vdots\\
x_N - y_N &= \int_0^t [g_N(x_1, x_2, \ldots, x_N) - g_N(y_1, y_2, \ldots, y_N)]\, dt_1.
\end{aligned}
$$

Let us now employ the multidimensional mean-value theorem. Essentially, we are asking that the g_i possess partial derivatives with respect to the x_i in some region containing the point $(c_1, c_2, \ldots, c_N)$. We obtain inequalities of the form

$$
(5.3)\qquad
\begin{aligned}
&|g_1(x_1, x_2, \ldots, x_N) - g_1(y_1, y_2, \ldots, y_N)|\\
&\qquad \le b[|x_1 - y_1| + |x_2 - y_2| + \cdots + |x_N - y_N|],\\
&|g_2(x_1, x_2, \ldots, x_N) - g_2(y_1, y_2, \ldots, y_N)|\\
&\qquad \le b[|x_1 - y_1| + |x_2 - y_2| + \cdots + |x_N - y_N|],\\
&\qquad \vdots\\
&|g_N(x_1, x_2, \ldots, x_N) - g_N(y_1, y_2, \ldots, y_N)|\\
&\qquad \le b[|x_1 - y_1| + |x_2 - y_2| + \cdots + |x_N - y_N|].
\end{aligned}
$$

where b denotes a common bound for the partial derivatives of g_2 with respect to its arguments. Let us use these inequalities in Eq. (5.2) and add the results. We obtain

$$
(5.4)\qquad
\begin{aligned}
&|x_1 - y_1| + |x_2 - y_2| + \cdots + |x_N - y_N|\\
&\qquad \le Nb \int_0^t [|x_1 - y_1| + |x_2 - y_2| + \cdots + |x_N - y_N|]\, dt_1.
\end{aligned}
$$

Using the Lemma of Section 6.2, we see that (5.4) implies that

$$|x_1 - y_1| + |x_2 - y_2| + \cdots + |x_N - y_N| \equiv 0. \tag{5.5}$$

Hence

$$x_1 \equiv y_1,\ x_2 \equiv y_2,\ \ldots,\ x_N \equiv y_N,$$

the desired uniqueness of the solution.

Consequently, we can assert

Theorem. If

$$g_1(x_1, x_2, \ldots, x_N), \ldots, g_N(x_1, x_2, \ldots, x_N)$$

are polynomials in $x_1, x_2, \ldots, x_N$, *then Eqs.* (5.1) *have at most one solution in any interval* $[0, t_0]$.

6.6 EXISTENCE

Let us now show how to establish the existence of a solution in a sufficiently small t-interval $[0, t_0]$. We shall employ the fundamental method of successive approximation. To illustrate this approach, let us begin with the equation

$$u' = u^2 + t, \qquad u(0) = 1. \tag{6.1}$$

Let $u_0(t) = 1$ be an initial guess. This is certainly a good approximation for small t. Then let us determine u_1, the next approximation, by means of the equation

$$u_1' = u_0^2 + t, \qquad u_1(0) = 1. \tag{6.2}$$

Continuing in this fashion, let us define

$$\begin{aligned} u_2' &= u_1^2 + t, \qquad & u_2(0) &= 1, \\ &\vdots \\ u_n' &= u_{n-1}^2 + t, \qquad & u_n(0) &= 1. \end{aligned} \tag{6.3}$$

Our objective is to show that the sequence of functions $\{u_n(t)\}$ converges uniformly to the solution of (6.1) as $n \to \infty$, provided that t_0 is sufficiently small. The proof proceeds in two stages. First, we show that $\{|u_n(t)|\}$ is uniformly bounded if t_0 is properly chosen; and then we establish convergence.

EXERCISES

1. Calculate $u_1(t)$, $u_2(t)$, $u_3(t)$, $u_4(t)$ explicitly as polynomials in t and plot their graphs in $0 \le t \le 1$ for the values $t = 0, 0.1, 0.2, \ldots, 0.9, 1.0$.
2. Compare the values obtained in this way with the power-series expansion of u truncated at the end of N terms for $N = 1, 2, 3, 4$. Which procedure yields a more accurate approximation over $0 \le t \le 1$?
3. Calculate the solution of $u' = u^2 + t$ using the algorithm

$$u(t + \Delta) = u(t) + \Delta[u(t)^2 + t], \qquad t = 0, \Delta, 2\Delta, \ldots, (N - 1)\Delta,$$

   where $N\Delta = 1$, for $\Delta = 0.1, 0.01$, and compare the results obtained with those of Exercises 1 and 2.
4. Consider the sequence $\{u_n(t)\}$ obtained using the equations

$$u'_{n+1} = 2u_n u_{n+1} - u_n^2 + t, \qquad u_{n+1}(0) = 1, \qquad n = 0, 1, \ldots,$$

   where $u_0 = 1$. Calculate the power-series expansions of the new sequence up to terms of order 10 for $n = 0, 1, 2, 3, 4$, and compare with the preceding results.
5. Using the foregoing recurrence relation, show that

$$1 \le u_1(t) \le u_2(t) \le \cdots \le u_n(t) \le \cdots \le u(t)$$

   for all t for which the solution $u(t)$ exists.
6. What happens if we use the method of successive approximations in the following way:

$$u_0 = 1, u_0' = u_1^2 + t, u_1' = u_2^2 + t, \ldots$$

6.7 BOUNDEDNESS

Let us suppose that we wish to find a value of t_0 such that $|u_n(t)| \le 2$ for $0 \le t \le t_0$ for $n = 0, 1, 2, \ldots$ We have

$$u_1 = 1 + \int_0^t (u_0^2 + t_1)\, dt_1, \tag{7.1}$$

$$|u_1| \le \left|1 + \int_0^t (u_0^2 + t_1)\, dt_1\right| \le 1 + \int_0^t |u_0^2 + t_1|\, dt_1$$

$$< 1 + \int_0^t (4 + t_0)\, dt_1 = 1 + (4 + t_0)t \le 1 + (4 + t_0)t_0.$$

Hence $|u_1| \leq 2$ if $1 + (4 + t_0)t_0 \leq 2$. This relation is certainly true if $t_0 \leq \frac{1}{5}$. We are not interested in obtaining the largest values of t_0 at this point, merely convenient ones. We see that this argument goes through inductively; namely, if $|u_n(t)| \leq 2$ for $0 \leq t \leq \frac{1}{5}$, then, using

$$u_{n+1} = 1 + \int_0^t (u_n^2 + t_1)\, dt_1, \tag{7.2}$$

we see that u_{n+1} satisfies the same bound.

EXERCISES

1. Show that $|u_n(t)| \leq \tan t$ for $n \geq 1$, $0 \leq t \leq 1$. [*Hint:*

$$|u_{n+1}| \leq \int_0^t [u_n^2 + 1]\, dt_1 \leq 1 + \int_0^t [\tan^2 t_1 + 1]\, dt_1, \quad \text{etc.}]$$

The point of this example is that with a little ingenuity and exploitation of the particular properties of the equation under consideration we can always improve the estimates obtained using the straightforward techniques of Section 6.7.

2. Recalling that the solution of $u' = u^2 + t$, $u(0) = 1$, can be made to depend on the solution of a second-order linear differential equation, determine the exact interval of existence of u to two significant figures.

6.8 CONVERGENCE

To prove that the sequence $\{u_n\}$ converges, we consider the infinite series

$$S = \sum_{n=0}^{\infty} (u_{n+1} - u_n). \tag{8.1}$$

Since the Nth partial sum is

$$\begin{aligned} S_N &= (u_1 - u_0) + (u_2 - u_1) + \cdots + (u_{N+1} - u_N) \\ &= u_{N+1} - u_0, \end{aligned} \tag{8.2}$$

we see that convergence of the series is equivalent to the convergence of the sequence $\{u_n\}$. The series converges if it converges absolutely, which is to say if the comparison series

$$\sum_{n=0}^{\infty} |u_{n+1} - u_n| \tag{8.3}$$

converges. Turning to Eq. (7.2) and subtracting the expressions for n from that for $n + 1$, we have

(8.4)
$$u_{n+1} - u_n = \int_0^t (u_n^2 - u_{n-1}^2)\,dt_1,$$
$$|u_{n+1} - u_n| \le \int_0^t |u_n^2 - u_{n-1}^2|\,dt_1 = \int_0^t |u_n - u_{n-1}| \cdot |u_n + u_{n-1}|\,dt_1$$
$$\le \int_0^t |u_n - u_{n-1}| \cdot \big|\,|u_n| + |u_{n-1}|\,\big|\,dt_1$$
$$\le 4\int_0^t |u_n - u_{n-1}|\,dt_1,$$

provided that $t_0 \le \frac{1}{5}$, using the bound on u_n established in Section 6.7. We have

(8.5)
$$u_1 - u_0 = u_1 - 1 = \int_0^t (u_0^2 + t_1)\,dt_1.$$

Hence

(8.6)
$$|u_1 - u_0| \le \int_0^t (1 + 1)\,dt_1 = 2t < 1 \qquad \text{for} \qquad t \in [0, 1/5].$$

Using (8.4), we have

(8.7)
$$|u_2 - u_1| \le 4\int_0^t |u_1 - u_0|\,dt_1 \le 4\int_0^t dt_1 = 4t,$$
$$|u_3 - u_2| \le 4\int_0^t |u_2 - u_1|\,dt_1 \le 4\int_0^t 4t_1\,dt_1 = \frac{16t^2}{2!}.$$

Inductively, we see that

(8.8)
$$|u_{n+1} - u_n| \le \frac{(4t)^n}{n!}.$$

Hence the series in (8.3) converges by comparison with the exponential series

(8.9)
$$\sum_{n=0}^{\infty} \frac{(4t)^n}{n!} = e^{4t}.$$

Observe that the success of our method depended on the use of the integral equation (7.2).

It remains to be shown that the limit function

(8.10)
$$u(t) = \lim_{n\to\infty} u_n(t)$$

satisfies the differential equation (6.1). Reviewing the foregoing proof of convergence, we see that we have established uniform convergence of $\{u_n(t)\}$ in $[0, t_0]$. This follows from the uniform convergence of the exponential series in $[0, t_0]$. Hence $u(t)$, as the uniform limit of a sequence of continuous functions, is continuous. That each $u_n(t)$ is continuous readily follows inductively.

Returning to Eq. (7.2), we see that we can pass to the limit on both sides by virtue of the uniform convergence, namely

$$
\begin{aligned}
(8.11) \qquad u(t) = \lim_{n\to\infty} u_{n+1}(t) &= 1 + \lim_{n\to\infty} \int_0^t (u_n^2 + t_1)\, dt_1 \\
&= 1 + \int_0^t \lim_{n\to\infty} (u_n^2 + t_1)\, dt_1 \\
&= 1 + \int_0^t (u^2 + t_1)\, dt_1.
\end{aligned}
$$

Since $u^2 + t$ is continuous in $[0, t_0]$, it follows that the derivative of $\int_0^t [u^2 + t_1]\, dt_1$ exists in this interval and is equal to $u^2 + t$. Hence, differentiating both sides of Eq. (8.11), we see that $u(t)$ satisfies the original differential equation.

EXERCISES

1. Using the estimate $|u_n(t)| \leq \tan t$ for $n \geq 1$, $0 \leq t \leq 1$, show that $\{u_n\}$ converges in the interval $0 \leq t \leq 1$.
2. Returning to Eqs. (6.3), show that the uniform convergence of $\{u_n\}$ in $[0, t_0]$ entails the uniform convergence of $\{u_n'\}$. Let

$$v(t) = \lim_{n\to\infty} u_n'(t).$$

Show that $v = u'$, so that u satisfies the original differential equation.

6.9 THE GENERAL CASE

To handle the equation

$$(9.1) \qquad u' = g(u, t), \qquad u(0) = c,$$

we require a technique for estimating the difference $g(u_n, t) - g(u_{n-1}, t)$. The Lipschitz condition imposed above in Section 6.3 is precisely what is needed.

The same approach is applicable to the study of systems of equations. As a matter of fact, if we use some simple vector notation, we can treat the one-dimensional case and the N-dimensional case in the same fashion.

EXERCISE

1. Consider the successive approximations

$$u_0 = c, \qquad u'_{n+1} = g(u_n, t), \qquad u_{n+1}(0) = c.$$

Construct a flow chart for a digital computer calculation of adaptive type in which the calculation continues until

$$\sum_{i=1}^{M} [u_{n+1}(t_i) - u_n(t_i)]^2 \leq \epsilon,$$

where the t_i are suitably chosen points in $[0, t_0]$.

6.10 TWO-POINT BOUNDARY-VALUE PROBLEMS

The method of successive approximations plays a particularly important role in the study of nonlinear differential equations subject to two-point boundary-value conditions. Thus, if we wished to obtain a computational solution of

$$u'' = u^2 + t, \qquad u(0) = 1, \qquad u(t_0) = 0, \tag{10.1}$$

we could begin by setting

$$u_0 = 1 - t/t_0 \tag{10.2}$$

[the straight-line approximation to a curve passing through (0, 1) and $(t_0, 0)$], and then continue as follows:

$$\begin{aligned} u_1'' &= u_0^2 + t, \qquad u_1(0) = 1, \qquad u_1(t_0) = 0,\\ &\vdots \\ u_n'' &= u_{n-1}^2 + t, \qquad u_n(0) = 1, \qquad u_n(t_0) = 0. \end{aligned} \tag{10.3}$$

Methods similar to those used in the foregoing sections can now be applied to establish the convergence of the sequence $\{u_n\}$ to the unique solution of Eq. (10.1), provided that t_0 is sufficiently small. The study of nonlinear two-point boundary-value problems is substantially more difficult than that of initial-value problems and belongs to a third course in differential equations.

6.11 QUASILINEARIZATION

The major importance of the foregoing method of successive approximations resides in the fact that the equation used to determine each new function u_n in terms of the preceding one is *linear.* This enables us to solve for u_{n+1} in terms of u_n in a simple fashion. We can preserve this linearity and considerably accelerate the convergence by using, in place of Eqs. (10.3), the approximations

$$\begin{aligned} u_1'' &= 2u_0u_1 - u_0^2 + t, \qquad u_1(0) = 1, \qquad u_1(t_0) = 0, \\ &\vdots \\ u_n'' &= 2u_{n-1}u_n - u_{n-1}^2 + t, \qquad u_n(0) = 1, \qquad u_n(t_0) = 0. \end{aligned} \tag{11.1}$$

The basic idea is simple and quite important in connection with numerical solutions. Instead of writing

$$g(u) \cong g(u_n), \tag{11.2}$$

we write

$$g(u) \cong g(u_n) + (u - u_n)g'(u_n), \tag{11.3}$$

which is a second-order rather than a first-order approximation. Thus $g(u) \cong g(u_n)$ implies that

$$|g(u) - g(u_n)| \leq b_1|u - u_n|$$

for some constant b_1, whereas (11.3) implies that

$$|g(u) - g(u_n) - (u - u_n)g'(u_n)| \leq b_2|u - u_n|^2.$$

The case $g(u) = u^2$ yields the recurrence relations (11.1).

It turns out that, using the relations (10.3), we can find an approximation of the type

$$|u - u_n| \leq ar^n, \tag{11.4}$$

for $0 \leq t \leq t_0$, $n \geq 0$, where $r < 1$. Using (11.1), we obtain an approximation of the type

$$|u - u_n| \leq a_1r^{2^n}, \tag{11.5}$$

for $0 \leq t \leq t_0, n \geq 0$. It would take us too far afield to discuss all the questions involved in obtaining these estimates. The difference between the rates of convergence of (11.4) and (11.5) becomes quite noticeable even for $n = 4$.

6.12 DISCUSSION OF THE CONVERGENCE OF FINITE DIFFERENCE ALGORITHMS

In Chapter 4, we described the use of difference equations such as

$$v_{k+1} - v_k = g(v_k)\Delta, \qquad v_0 = c, \tag{12.1}$$

in obtaining numerical values of the solution of the differential equation

$$u' = g(u), \qquad u(0) = c. \tag{12.2}$$

Here v_k is an approximate value of $u(k\Delta)$, $k = 0, 1, \ldots, N$, where $N\Delta = t_0$, and it is supposed that the solution of (12.2) exists and is unique in $[0, t_0]$. As we saw before, for this to be true it is sufficient that $g(u)$ have a derivative which is bounded for u in the neighborhood of $u = c$. Presumably the smaller the value of Δ, the more accurate is the approximation to the solution of Eq. (12.1). This is the result we wish to demonstrate in this section.

Before proceeding with the proof, let us point out that there are two distinct uses for difference equations. The first is that of establishing the existence of a solution from the beginning; the second is that of obtaining approximate values for the solution once we have proved in some other manner that it exists. The first task is considerably more difficult than the second. We shall consider only the latter here.

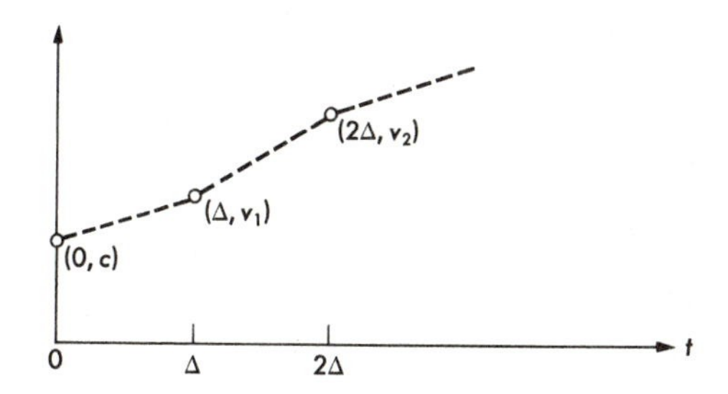

Figure 6.1

Using Eq. (12.1), we obtain a sequence of values, $v_0, v_1, \ldots, v_N$. To obtain a function defined for all t in $[0, t_0]$, we use straight-line interpolation to fill in the missing values. Geometrically, the procedure is illustrated in Fig. 6.1.

Analytically, we have

$$w(t) = v_k + \frac{(t - k\Delta)(v_{k+1} - v_k)}{\Delta} \tag{12.3}$$

for $k\Delta \leq t \leq (k + 1)\Delta$, $k = 0, 1, \ldots, (N - 1)$. Since this function $w(t)$ clearly depends on Δ, let us write $w(t, \Delta)$. What we wish to demonstrate is that

$$\lim_{\Delta \to 0} w(t, \Delta) = u(t) \tag{12.4}$$

uniformly for $0 \leq t \leq t_0$.

There are, as usual, several ways to proceed. Let us show that we can interpret the problem as one of stability. From (12.2) and integrating between $t = k\Delta$ and $t = (k + 1)\Delta$, we have

$$\begin{aligned} u((k+1)\Delta) - u(k\Delta) &= \int_{k\Delta}^{(k+1)\Delta} u'\,dt_1 \\ &= \int_{k\Delta}^{(k+1)\Delta} g(u)\,dt_1. \end{aligned} \tag{12.5}$$

Regarding $g(u)$ as a function of t and using the mean-value theorem, we obtain

$$g[u(t)] = g[u(k\Delta)] + (t - k\Delta)g'[u(\theta)]u'(\theta) \tag{12.6}$$

for $k\Delta \leq t \leq (k + 1)\Delta$, where

$$k\Delta \leq \theta \leq (k + 1)\Delta.$$

Here $\theta = \theta(t)$. Since Eq. (12.2) holds, we have

$$|g'[u(\theta)]u'(\theta)| = |g'[u(\theta)]g[u(\theta)]| \leq b_1, \tag{12.7}$$

where b_1 is the maximum value of $|g'[u(t)]g[u(t)]|$ for $0 \leq t \leq t_0$. Thus we see that

(12.8)

$$\begin{aligned} \int_{k\Delta}^{(k+1)\Delta} g(u)\,dt_1 &= \int_{k\Delta}^{(k+1)\Delta} g[u(k\Delta)]\,dt_1 + \int_{k\Delta}^{(k+1)\Delta} (t_1 - k\Delta)g'[u(\theta)]u'(\theta)\,dt_1 \\ &= g[u(k\Delta)]\Delta + \epsilon_k, \end{aligned}$$

where

$$(12.9)\qquad |\epsilon_k| = \left|\int_{k\Delta}^{(k+1)\Delta} (t_1 - k\Delta)g'[u(\theta)]u'(\theta)\, dt_1\right|$$

$$\leq b_1 \int_{k\Delta}^{(k+1)\Delta} (t_1 - k\Delta)\, dt_1$$

$$= \frac{b_1 \Delta^2}{2}.$$

Writing $u_k = u(k\Delta)$, we see then that (12.5) leads to

$$(12.10)\qquad u_{k+1} - u_k = g(u_k)\Delta + \epsilon_k,$$

$$u_0 = c,$$

where (12.9) holds.

Consequently, as mentioned above, the problem is one of stability. We are required to show that the small perturbing term ϵ_k exerts no appreciable influence on the sequence $\{u_k\}$ generated by (12.10).

In particular what we wish to demonstrate is that

$$(12.11)\qquad |u_k - v_k| \leq b_2\Delta$$

for $k = 0, 1, \ldots, N$, where b_2 is a constant, independent of both k and Δ.

Although the proof of this result is not difficult in the sense of requiring any more analysis than the reader has at his disposal at this point, it does require a sustained use of arithmetic and algebraic estimates. Consequently, we feel that the proof lies outside the agreed-upon province of the book.

The reader interested in the details and in learning more about this vital domain of modern mathematics will find references at the end of the chapter. The ambitious reader is urged to establish (12.11) in his own fashion.

BIBLIOGRAPHY AND COMMENTS

Section 6.1. For more detailed treatments of existence and uniqueness of solutions of differential equations, see:

BELLMAN, R., *Stability Theory of Differential Equations*, Dover Publications, New York, 1969.

CODDINGTON, E. A., and N. LEVINSON, *Theory of Ordinary Differential Equations*, McGraw-Hill, New York, 1955.

HARTMAN, P., *Ordinary Differential Equations*, Wiley, New York, 1964.

Section 6.10. For a more detailed discussion and many further references on boundary-value problems, see:

BELLMAN, R., and R. KALABA, *Quasilinearization and Nonlinear Boundary Value Problems*, American Elsevier, New York, 1965.

Answers to Selected Exercises

Section 1.5

1. $\dfrac{v}{g} \pm \dfrac{(v^2 - 2gh)^{1/2}}{g}$

3. $\dfrac{v}{g}$; $\dfrac{v}{g} + \dfrac{(v^2 + 2gh)^{1/2}}{g}$

Section 1.8

1. $\log N_0 = \dfrac{(\sum t_i^2)(\sum \log E(t_i)) - (\sum t_i)(\sum t_i \log E(t_i))}{N\sum t_i^2 - (\sum t_i)^2}$

$$k = \frac{N\sum t_i \log E(t_i) - (\sum t_i)(\sum \log E(t_i))}{N\sum t_i^2 - (\sum t_i)^2}$$

Section 1.11

3. If the light source is placed at the origin of coordinates, the equation is $y = 2xy' + y(y')^2$.

6. $T = \dfrac{1}{\sqrt{2g}} \displaystyle\int_0^x \dfrac{s'(u)\,du}{\sqrt{x - u}}$, where $s(u)$ is arc length along the curve.

Section 1.13

1. $u'(t) + (1 - a)u(t) + ak = 0, \quad u(0) = k$

3. $u''(x) - (a^2 + 2ab)u(x) = -a^2k,$
   $au(0) - u'(0) = ak,$
   $au(1) + u'(1) = ak.$

Section 1.17

1. Since u is linear, two pairs of observed values of t and u determine the constants c, k. If more than two observations are made, c and k can be estimated by least squares fit. (See Section 4.18.)

Section 1.21

1. $u = 1/b$
3. If $u(t_1) = t_1$, then $u'(t_1) = 0$ and $u''(t_1) = 1 - u'(t_1) > 0$.
5. If $u(t_1) = u(t_2) = 0$, $t_1 < t_2$, there must be a positive maximum or negative minimum between t_1 and t_2. At this extremum, $u'' = gu$ is positive in the first case, negative in the second, which is impossible.

Section 1.22

1. No

Section 2.3

4. For $b \neq -a$, $y = \dfrac{(e^{bx} - e^{-ax})}{(a + b) + y(0)e^{-ax}}$

   For $b = -a$, $y = xe^{-ax} + y(0)e^{-ax}$

8. $y = \left(c - \dfrac{m}{a}\right)e^{-ax} + \dfrac{m}{a}$, where $m = \dfrac{ac(e^{-ab} - 1)}{a(b - a) - 1 + e^{-ab}}$

13. $u(x) = g(x) - e^{-x}\int_0^x e^{x_1}g(x_1)\,dx_1$

Section 2.4

3. $y = Ax^2 + Bx + C$ yields $A = 0$, $B = 1$, $C = -1$.
4. $y = c\exp\left[-\int_0^x a(u)\,du\right] \Big/ \int_0^b \exp\left[-\int_0^{x_1} a(u)\,du\right] dx_1$
7. Let $v(x) = \int_0^x e^{-ax_1}u(x_1)\,dx_1$.

Section 2.13

1. (a) $z = \dfrac{A\cos - B\sin x}{A\sin x + B\cos x}$ (b) $z = \dfrac{Ae^x - Be^{-x}}{Ae^x + Be^{-x}}$

   (c) $z = \dfrac{Ar_1e^{r_1x} + Br_2e^{r_2x}}{Ae^{r_1x} + Be^{r_2x}}$, $r_1 = \dfrac{-1 + \sqrt{5}}{2}$, $r_2 = \dfrac{-1 - \sqrt{5}}{2}$

4. If $a^2 - 4b \neq 0$, all solutions have the form

$$y = c_1e^{r_1x} + c_2e^{r_2x},$$

   where $r_1, r_2 = (-a \pm \sqrt{a^2 - 4b})/2$. Consider $a^2 > 4b$ and $a^2 < 4b$ separately.

5. Since $y'y'' + yy' = \frac{1}{2}\dfrac{d}{dx}[(y')^2 + y^2] = 0$, it follows that

$$(y')^2 + y^2 = \text{constant} = (y'(0))^2 + y(0)^2 = 1.$$

Section 2.15

1. $t = RC \log 2$

Section 2.18

1. $u_{ss} = \dfrac{E_0}{L}\left[\dfrac{R/L \cos \omega t + \omega \sin \omega t}{(R/L)^2 + \omega^2}\right]$

Section 2.22

1. $u = \frac{2}{3}e^{-t} + \frac{1}{3}e^{2t}$

Section 2.23

1. a. $y = c_1 + c_2e^{-t} - (2 \sin 2t + \cos 2t)/10$
   e. $y = c_1e^{at} + c_2e^{-at} - (\omega^2 + a^2)^{-1} \sin \omega t$

Section 2.29

2. $a_1 = a + 2k$, $b_1 = k^2 + ak + b$. Hence a_1 and b_1 are positive if k is large.

7. $v'' + \left[b - \dfrac{a^2}{4}\right]v = 0$

1. $y = e^{rx}\displaystyle\int_0^x \left[\int_0^{x_1} e^{-rx_2}g(x_2)\, dx_2\right] dx_1$

2. $y = c_1e^{r_1t} + c_2e^{r_2t} + \dfrac{e^{r_2t} - r_2t - 1}{(r_2 - r_1)r_2^2} - \dfrac{e^{r_1t} - r_1t - 1}{(r_2 - r_1)r_1^2}$,

   where

$$r_1 = -\frac{a}{2} + \sqrt{\frac{a^2}{4} - b}, \qquad r_2 = -\frac{a}{2} - \sqrt{\frac{a^2}{4} - b}.$$

5. $u = 1 - \cos t, \quad 0 \le t \le \pi; \qquad u = -2 \cos t, \quad \pi < t$

Section 2.31

1. $u = a_1 \cos px + a_2 \sin px$, where

$$a_1 = \frac{c_1(\sin pa - ap \cos pa + ap) - c_2p^2(1 - \cos pa)}{2(1 - \cos pa)},$$

$$a_2 = \frac{-c_1(\cos pa + ap \sin pa - 1) + c_2p^2 \sin pa}{2(1 - \cos pa)}$$

5. When $\sin ba \neq 0$.

Section 2.33

1. $y = 1 + 2 \log x$

Section 2.34

7. $u_n = \dfrac{s^n}{s^2 + as + b}; \quad u_n = \dfrac{ns^{n-1}}{2s + a}$

11. $u_n = \left(\prod_{k=0}^{n-1} a_k\right) c + \left(\prod_{k=1}^{n-1} a_k\right) f_0 + \cdots + a_{n-1}f_{n-2} + f_{n-1}$

13. $\lim_{n\to\infty} u_n = 2$

14. The solution must have the form (34.5), where

$$a_1 + a_2 = c_1, \qquad a_1 r_1^N + a_2 r_2^N = c_2.$$

Section 3.3

1. $f^2(x) = \sum_{n=0}^{\infty}\left[\sum_{i=0}^{n} a_i a_{n-i}\right] x^n$

2. $$e^f = 1 + a_0 + \tfrac{1}{2}a_0^2 + \tfrac{1}{6}a_0^3 + \cdots + (a_1 + a_0a_1 + \tfrac{1}{2}a_0^2a_1 + \cdots)x$$
$$+ (a_2 + \tfrac{1}{2}a_1^2 + a_0a_2 + \tfrac{1}{2}a_0a_1^2 + \tfrac{1}{2}a_0^2a_2 + \cdots)x^2$$
$$+ (a_3 + a_0a_3 + a_1a_2 + \tfrac{1}{6}a_1^3 + \tfrac{1}{2}a_0^2a_3 + a_0a_1a_2 + \cdots)x^3 + \cdots$$

$$\sin f = a_0 - \tfrac{1}{6}a_0^3 + \tfrac{1}{120}a_0^5 - \cdots$$
$$+ (a_1 - \tfrac{1}{2}a_0^2a_1 + \tfrac{1}{24}a_0^4a_1 + \cdots)x$$
$$+ (a_2 - \tfrac{1}{2}a_0a_1^2 - \tfrac{1}{2}a_0^2a_2 + \tfrac{1}{12}a_0^3a_1^2 + \tfrac{1}{24}a_0^4a_2 + \cdots)x^2$$
$$+ (a_3 - \tfrac{1}{6}a_1^3 - \tfrac{1}{2}a_0^2a_3 - a_0a_1a_2 - \cdots)x^3 + \cdots$$

Section 3.4

1. $$e^x = 1 + \frac{x}{1!} + \frac{x^2}{2!} + \frac{x^3}{3!} + \cdots$$
$$\cos x = 1 - \frac{x^2}{2!} + \frac{x^4}{4!} - \frac{x^6}{6!} + \cdots$$
$$\sin x = x - \frac{x^3}{3!} + \frac{x^5}{5!} - \frac{x^7}{7!} + \cdots$$

9. $x_2 = \frac{8}{7} \cong 1.4286, \quad x_3 = \frac{1203}{1043} \cong 1.1534$

Section 3.7

5. $u = e^{x^2}\int_0^x e^{-x_1^2}\,dx_1 = x + \frac{2}{3}x^3 + \frac{4}{15}x^5 + \frac{8}{105}x^7 + \cdots$

Section 3.8

2. (a) $$u_1(x) = 1 - \frac{n(n+1)}{2!}x^2 + \frac{n(n-2)(n+1)(n+3)}{4!}x^4$$
$$- \frac{n(n-2)(n-4)(n+1)(n+3)(n+5)}{6!}x^6 + \cdots$$

$$u_2(x) = x - \frac{(n-1)(n+2)}{3!}x^3 + \frac{(n-1)(n-3)(n+2)(n+4)}{5!}x^5$$
$$- \frac{(n-1)(n-3)(n-5)(n+2)(n+4)(n+6)}{7!}x^7 + \cdots$$

(b) $n = 0, u = 1; n = 1, u = x; n = 2, u = 1 - 3x^2; n = 3, u = x - \frac{5}{3}x^3$.

Section 3.10

2. (a) $$u = \sin x = \frac{1}{\sqrt{2}}\left[1 + \left(x - \frac{\pi}{4}\right) - \frac{(x - \pi/4)^2}{2!} - \frac{(x - \pi/4)^3}{3!} + \cdots\right],$$

$$u = \cos x = \frac{1}{\sqrt{2}}\left[1 - \left(x - \frac{\pi}{4}\right) - \frac{(x - \pi/4)^2}{2!} + \frac{(x - \pi/4)^3}{3!} + \cdots\right].$$

(b) $u_1(x) = 1 + \frac{1}{2}(x - 1)^2 + \frac{1}{6}(x - 1)^3 + \frac{1}{6}(x - 1)^4 + \cdots,$
$u_2(x) = (x - 1) + \frac{1}{2}(x - 1)^2 + \frac{1}{2}(x - 1)^3 + \frac{1}{4}(x - 1)^4 + \cdots.$

Section 3.11

1. $u = \tan x = x + \frac{1}{3}x^3 + \frac{2}{15}x^5 + \frac{17}{315}x^7 + \cdots$
3. $u = \log \cos x = -\frac{1}{2}x^2 - \frac{1}{12}x^4 - \frac{1}{45}x^6 - \frac{17}{2520}x^8 - \cdots$

Section 3.12

4. The error in stopping with x^n/n is at most $1/(n + 1)$. Hence the error is less than 0.01 if $n \geqq 100$.

Section 3.15

1. $u = \frac{1}{2} + \frac{2}{5}x + \frac{62}{125}x^2 - \frac{408}{1875}x^3 + \cdots$
$u = 1 + 2x + 2x^2 + 4x^3 + \cdots$

Section 3.18

5. $$y = J_{-1}(x) = -\frac{x}{2} + \frac{x^3}{2^3 1!1!} - \frac{x^5}{2^5 2!3!} + \cdots$$

Section 3.21

1. $$u_2(x) = \tfrac{1}{8}\sin x(\tfrac{5}{4}x^2 \sin 2x - \tfrac{7}{8}\sin 2x - \tfrac{1}{2}x^3 \cos 2x + \tfrac{7}{4}x \cos 2x + \tfrac{1}{3}x^3) + \tfrac{1}{8}\cos x(\tfrac{5}{4}x^2 \cos 2x - \tfrac{7}{8}\sin 2x + \tfrac{1}{2}x^3 \cos 2x - \tfrac{7}{4}x \sin 2x - \tfrac{1}{4}x^4 + \tfrac{1}{2}x^2 + \tfrac{7}{8})$$

2. Equal. $$u = 1 - \frac{\epsilon x^2}{2!} + \frac{\epsilon^2 x^4}{4!} - \cdots$$

Section 3.24

2. $u_1 = -\frac{1}{4}\sin^5 x - \frac{1}{4}\cos^2 x \sin^3 x - \frac{3}{8}\cos^2 x \sin x + \frac{3}{8}x \cos x$

Miscellaneous Exercises

2. Since $\sum_{n=0}^{\infty} (a_n/b^n)$ converges, there is a k such that $a_n \leqq kb^n$ for $n = 0, 1, 2, \ldots$
4. $0 \leq a_n \leq b_n$ implies $0 \leq na_n \leq nb_n$.
18. $-a_1 x w_1' = 1 - 3a_1 - a_1^2 x^2 + (2 - 3a_1)w_1 + w_1^2$

Section 4.3

3. $\log(u + \sqrt{u^2 + 1}) - \log(u_0 + \sqrt{u_0^2 + 1}) = -\frac{1}{n}\log(y/y_0)$

4. $\frac{dx}{dy} = \frac{c}{2}y^{-1/n} - \frac{1}{2c}y^{1/n}, \quad c = y_0^{1/n}(u_0 + \sqrt{u_0^2 + 1})$

   For $n = 1$,
$$x = \frac{c}{2}\log y - \frac{1}{4c}y^2.$$
   For $n > 1$,
$$x = \frac{c}{2(1 - 1/n)}y^{1-(1/n)} - \frac{1}{2c(1 + 1/n)}y^{1+(1/n)}.$$

Section 4.5

3. The last few values are:

Time	$u_1(t)$	$u_2(t)$
0.8	0.44991481	0.44933495
0.9	0.40719495	0.40657575
1.0	0.36847582	0.36788557

4. $u_n = b_1(-\Delta + \sqrt{\Delta^2 + 1})^n + b_2(-\Delta - \sqrt{\Delta^2 + 1})^n$, where
$$b_1 = \frac{1 + (\Delta^2/2) + R}{2R}, \qquad b_2 = \frac{-1 - (\Delta^2/2) + R}{2R},$$
$$R = \sqrt{\Delta^2 + 1}.$$
   The computed values agree with those in Exercise 3.

Section 4.10

3. 0.680068 (extrapolated value), 0.739068 (computed value)

4. $u = 0.1 + 0.01t + 0.501t^2 + 0.03343333t^3 + 0.00417667t^4 + 0.05050100t^5 + \ldots$

t	$u(t)$
0.1	0.106044
0.2	0.122330
0.3	0.149149
0.4	0.186923

Section 4.13

1.

t	$u(t)$
0.2	0.12224905
0.4	0.18669121
0.6	0.29783074
0.8	0.46576769
1.0	0.71238397

Section 5.7

1. (a) $u = 2e^{-t} + t - 1$ (c) $u = t$

(e) $$u = \left(\frac{1}{2} - \frac{5}{2\sqrt{21}}\right)\cos\left(\frac{5}{2} + \frac{\sqrt{21}}{2}\right)^{1/2} t$$
$$+ \left(\frac{1}{2} - \frac{5}{2\sqrt{21}}\right)\left(\frac{5}{2} + \frac{\sqrt{21}}{2}\right)^{-1/2} \sin\left(\frac{5}{2} + \frac{\sqrt{21}}{2}\right)^{1/2} t$$
$$+ \left(\frac{1}{2} + \frac{5}{2\sqrt{21}}\right)\cos\left(\frac{5}{2} - \frac{\sqrt{21}}{2}\right)^{1/2} t$$
$$+ \left(\frac{1}{2} + \frac{5}{2\sqrt{21}}\right)\left(\frac{5}{2} - \frac{\sqrt{21}}{2}\right)^{-1/2} \sin\left(\frac{5}{2} - \frac{\sqrt{21}}{2}\right)^{1/2} t$$

Section 5.9

1. $x_1 = x_2 = x_3 = e^{4t}$

Miscellaneous Exercises

3. $$u = c_1 \frac{r_1 e^{r_1 t} - r_2 e^{r_2 t}}{r_1 - r_2} + (c_2 + a_1 c_1)\frac{e^{r_1 t} - e^{r_2 t}}{r_1 - r_2}$$
$$+ \int_0^t \frac{e^{r_1(t-t_1)} - e^{r_2(t-t_1)}}{r_1 - r_2} f(t_1)\, dt_1,$$
where r_1, r_2 are the roots (assumed distinct) of $r^2 + a_1 r + a_2 = 0$.

5. The general form of $v(x, s)$ is
$$c_1 e^{\sqrt{s}x} + c_2 e^{-\sqrt{s}x} - \int_0^x \frac{e^{\sqrt{s}(x-x_1)} - e^{-\sqrt{s}(x-x_1)}}{2\sqrt{s}} g(x_1)\, dx_1.$$

19. $$a_1 = \frac{u_1 u_2'' - u_1'' u_2}{u_1' u_2 - u_1 u_2'} \qquad a_2 = \frac{u_1'' u_2' - u_1' u_2''}{u_1' u_2 - u_1 u_2'}$$

Section 6.6

1. $u_1 = 1 + t + \frac{1}{2}t^2,$
$u_2 = 1 + t + \frac{3}{2}t^2 + \frac{2}{3}t^3 + \frac{1}{4}t^4 + \frac{1}{20}t^5,$
$u_3 = 1 + t + \frac{3}{2}t^2 + \frac{4}{3}t^3 + \frac{13}{12}t^4 + \frac{49}{60}t^5 + \frac{13}{30}t^6$
$+ \frac{233}{1260}t^7 + \frac{29}{480}t^8 + \frac{31}{2160}t^9 + \frac{1}{400}t^{10} + \frac{1}{4400}t^{11}.$

2. $u = 1 + t + \frac{3}{2}t^2 + \frac{4}{3}t^3 + \frac{17}{12}t^4 + \cdots$

Section 6.7

2. (0, 0.93)

Author Index

Subject Index

A CATALOG OF SELECTED

DOVER BOOKS

IN SCIENCE AND MATHEMATICS

ASYMPTOTIC METHODS IN ANALYSIS, N.G. de Bruijn. An inexpensive, comprehensive guide to asymptotic methods–the pioneering work that teaches by explaining worked examples in detail. Index. 224pp. 5⅜ x 8½. 64221-6 Pa. $7.95

OPTICAL RESONANCE AND TWO-LEVEL ATOMS, L. Allen and J. H. Eberly. Clear, comprehensive introduction to basic principles behind all quantum optical resonance phenomena. 53 illustrations. Preface. Index. 256pp. 5⅜ x 8½. 65533-4 Pa. $10.95

COMPLEX VARIABLES, Francis J. Flanigan. Unusual approach, delaying complex algebra till harmonic functions have been analyzed from real variable viewpoint. Includes problems with answers. 364pp. 5⅜ x 8½. 61388-7 Pa. $10.95

ATOMIC SPECTRA AND ATOMIC STRUCTURE, Gerhard Herzberg. One of best introductions; especially for specialist in other fields. Treatment is physical rather than mathematical. 80 illustrations. 257pp. 5⅜ x 8½. 60115-3 Pa. $7.95

APPLIED COMPLEX VARIABLES, John W. Dettman. Step-by-step coverage of fundamentals of analytic function theory–plus lucid exposition of five important applications: Potential Theory; Ordinary Differential Equations; Fourier Transforms; Laplace Transforms; Asymptotic Expansions. 66 figures. Exercises at chapter ends. 512pp. 5⅜ x 8½. 64670-X Pa. $14.95

ULTRASONIC ABSORPTION: An Introduction to the Theory of Sound Absorption and Dispersion in Gases, Liquids and Solids, A.B. Bhatia. Standard reference in the field provides a clear, systematically organized introductory review of fundamental concepts for advanced graduate students, research workers. Numerous diagrams. Bibliography. 440pp. 5⅜ x 8½. 64917-2 Pa. $11.95

UNBOUNDED LINEAR OPERATORS: Theory and Applications, Seymour Goldberg. Classic presents systematic treatment of the theory of unbounded linear operators in normed linear spaces with applications to differential equations. Bibliography. 199pp. 5⅜ x 8½. 64830-3 Pa. $7.95

LIGHT SCATTERING BY SMALL PARTICLES, H.C. van de Hulst. Comprehensive treatment including full range of useful approximation methods for researchers in chemistry, meteorology and astronomy. 44 illustrations. 470pp. 5⅜ x 8½. 64228-3 Pa. $12.95

CONFORMAL MAPPING ON RIEMANN SURFACES, Harvey Cohn. Lucid, insightful book presents ideal coverage of subject. 334 exercises make book perfect for self-study. 55 figures. 352pp. 5⅝ x 8¼. 64025-6 Pa. $11.95

OPTICKS, Sir Isaac Newton. Newton's own experiments with spectroscopy, colors, lenses, reflection, refraction, etc., in language the layman can follow. Foreword by Albert Einstein. 532pp. 5⅜ x 8½. 60205-2 Pa. $13.95

GENERALIZED INTEGRAL TRANSFORMATIONS, A.H. Zemanian. Graduate-level study of recent generalizations of the Laplace, Mellin, Hankel, K. Weierstrass, convolution and other simple transformations. Bibliography. 320pp. 5⅜ x 8½. 65375-7 Pa. $8.95

CATALYSIS IN CHEMISTRY AND ENZYMOLOGY, William P. Jencks. Exceptionally clear coverage of mechanisms for catalysis, forces in aqueous solution, carbonyl- and acyl-group reactions, practical kinetics, more. 864pp. 5⅜ x 8½. 65460-5 Pa. $19.95

PROBABILITY: An Introduction, Samuel Goldberg. Excellent basic text covers set theory, probability theory for finite sample spaces, binomial theorem, much more. 360 problems. Bibliographies. 322pp. 5⅜ x 8½. 65252-1 Pa. $10.95

LIGHTNING, Martin A. Uman. Revised, updated edition of classic work on the physics of lightning. Phenomena, terminology, measurement, photography, spectroscopy, thunder, more. Reviews recent research. Bibliography. Indices. 320pp. 5⅝ x 8¼. 64575-4 Pa. $8.95

PROBABILITY THEORY: A Concise Course, Y.A. Rozanov. Highly readable, self-contained introduction covers combination of events, dependent events, Bernoulli trials, etc. Translation by Richard Silverman. 148pp. 5⅝ x 8¼. 63544-9 Pa. $8.95

AN INTRODUCTION TO HAMILTONIAN OPTICS, H. A. Buchdahl. Detailed account of the Hamiltonian treatment of aberration theory in geometrical optics. Many classes of optical systems defined in terms of the symmetries they possess. Problems with detailed solutions. 1970 edition. xv + 360pp. 5⅜ x 8½. 67597-1 Pa. $10.95

STATISTICS MANUAL, Edwin L. Crow, et al. Comprehensive, practical collection of classical and modern methods prepared by U.S. Naval Ordnance Test Station. Stress on use. Basics of statistics assumed. 288pp. 5⅜ x 8½. 60599-X Pa. $8.95

DICTIONARY/OUTLINE OF BASIC STATISTICS, John E. Freund and Frank J. Williams. A clear concise dictionary of over 1,000 statistical terms and an outline of statistical formulas covering probability, nonparametric tests, much more. 208pp. 5⅜ x 8½. 66796-0 Pa. $7.95

STATISTICAL METHOD FROM THE VIEWPOINT OF QUALITY CONTROL, Walter A. Shewhart. Important text explains regulation of variables, uses of statistical control to achieve quality control in industry, agriculture, other areas. 192pp. 5⅜ x 8½. 65232-7 Pa. $8.95

METHODS OF THERMODYNAMICS, Howard Reiss. Outstanding text focuses on physical technique of thermodynamics, typical problem areas of understanding, and significance and use of thermodynamic potential. 1965 edition. 238pp. 5⅜ x 8½. 69445-3 Pa. $8.95

STATISTICAL ADJUSTMENT OF DATA, W. Edwards Deming. Introduction to basic concepts of statistics, curve fitting, least squares solution, conditions without parameter, conditions containing parameters. 26 exercises worked out. 271pp. 5⅜ x 8½. 64685-8 Pa. $9.95

TENSOR CALCULUS, J.L. Synge and A. Schild. Widely used introductory text covers spaces and tensors, basic operations in Riemannian space, non-Riemannian spaces, etc. 324pp. 5⅝ x 8¼. 63612-7 Pa. $11.95

ROTARY-WING AERODYNAMICS, W.Z. Stepniewski. Clear, concise text covers aerodynamic phenomena of the rotor and offers guidelines for helicopter performance evaluation. Originally prepared for NASA. 537 figures. 640pp. 6⅛ x 9¼.
64647-5 Pa. $16.95

DIFFERENTIAL GEOMETRY, Heinrich W. Guggenheimer. Local differential geometry as an application of advanced calculus and linear algebra. Curvature, transformation groups, surfaces, more. Exercises. 62 figures. 378pp. 5⅜ x 8½.
63433-7 Pa. $11.95

INTRODUCTION TO SPACE DYNAMICS, William Tyrrell Thomson. Comprehensive, classic introduction to space-flight engineering for advanced undergraduate and graduate students. Includes vector algebra, kinematics, transformation of coordinates. Bibliography. Index. 352pp. 5⅜ x 8½. 65113-4 Pa. $10.95

THE THEORY OF GROUPS, Hans J. Zassenhaus. Well-written graduate-level text acquaints reader with group-theoretic methods and demonstrates their usefulness in mathematics. Axioms, the calculus of complexes, homomorphic mapping, *p*-group theory, more. Many proofs shorter and more transparent than older ones. 276pp. 5⅜ x 8½. 40922-8 Pa. $12.95

ANALYTICAL MECHANICS OF GEARS, Earle Buckingham. Indispensable reference for modern gear manufacture covers conjugate gear-tooth action, gear-tooth profiles of various gears, many other topics. 263 figures. 102 tables. 546pp. 5⅜ x 8½.
65712-4 Pa. $16.95

SET THEORY AND LOGIC, Robert R. Stoll. Lucid introduction to unified theory of mathematical concepts. Set theory and logic seen as tools for conceptual understanding of real number system. 496pp. 5⅜ x 8¼. 63829-4 Pa. $14.95

A HISTORY OF MECHANICS, René Dugas. Monumental study of mechanical principles from antiquity to quantum mechanics. Contributions of ancient Greeks, Galileo, Leonardo, Kepler, Lagrange, many others. 671pp. 5⅜ x 8½.
65632-2 Pa. $18.95

FAMOUS PROBLEMS OF GEOMETRY AND HOW TO SOLVE THEM, Benjamin Bold. Squaring the circle, trisecting the angle, duplicating the cube: learn their history, why they are impossible to solve, then solve them yourself. 128pp. 5⅜ x 8½. 24297-8 Pa. $5.95

MECHANICAL VIBRATIONS, J.P. Den Hartog. Classic textbook offers lucid explanations and illustrative models, applying theories of vibrations to a variety of practical industrial engineering problems. Numerous figures. 233 problems, solutions. Appendix. Index. Preface. 436pp. 5⅜ x 8½. 64785-4 Pa. $13.95

CURVATURE AND HOMOLOGY: Enlarged Edition, Samuel I. Goldberg. Revised edition examines topology of differentiable manifolds; curvature, homology of Riemannian manifolds; compact Lie groups; complex manifolds; curvature, homology of Kaehler manifolds. New Preface. Four new appendixes. 416pp. 5⅜ x 8½.
40207-X Pa. $14.95

HISTORY OF STRENGTH OF MATERIALS, Stephen P. Timoshenko. Excellent historical survey of the strength of materials with many references to the theories of elasticity and structure. 245 figures. 452pp. 5⅜ x 8½. 61187-6 Pa. $14.95

SPECIAL FUNCTIONS, N.N. Lebedev. Translated by Richard Silverman. Famous Russian work treating more important special functions, with applications to specific problems of physics and engineering. 38 figures. 308pp. 5⅜ x 8½. 60624-4 Pa. $9.95

THE EXTRATERRESTRIAL LIFE DEBATE, 1750–1900, Michael J. Crowe. First detailed, scholarly study in English of the many ideas that developed between 1750 and 1900 regarding the existence of intelligent extraterrestrial life. Examines ideas of Kant, Herschel, Voltaire, Percival Lowell, many other scientists and thinkers. 16 illustrations. 704pp. 5⅜ x 8½. 40675-X Pa. $19.95

INTEGRAL EQUATIONS, F.G. Tricomi. Authoritative, well-written treatment of extremely useful mathematical tool with wide applications. Volterra Equations, Fredholm Equations, much more. Advanced undergraduate to graduate level. Exercises. Bibliography. 238pp. 5⅜ x 8½. 64828-1 Pa. $8.95

POPULAR LECTURES ON MATHEMATICAL LOGIC, Hao Wang. Noted logician's lucid treatment of historical developments, set theory, model theory, recursion theory and constructivism, proof theory, more. 3 appendixes. Bibliography. 1981 edition. ix + 283pp. 5⅜ x 8½. 67632-3 Pa. $10.95

MODERN NONLINEAR EQUATIONS, Thomas L. Saaty. Emphasizes practical solution of problems; covers seven types of equations. ". . . a welcome contribution to the existing literature...."–*Math Reviews*. 490pp. 5⅜ x 8½. 64232-1 Pa. $13.95

FUNDAMENTALS OF ASTRODYNAMICS, Roger Bate et al. Modern approach developed by U.S. Air Force Academy. Designed as a first course. Problems, exercises. Numerous illustrations. 455pp. 5⅜ x 8½. 60061-0 Pa. $12.95

INTRODUCTION TO LINEAR ALGEBRA AND DIFFERENTIAL EQUATIONS, John W. Dettman. Excellent text covers complex numbers, determinants, orthonormal bases, Laplace transforms, much more. Exercises with solutions. Undergraduate level. 416pp. 5⅜ x 8½. 65191-6 Pa. $11.95

INCOMPRESSIBLE AERODYNAMICS, edited by Bryan Thwaites. Covers theoretical and experimental treatment of the uniform flow of air and viscous fluids past two-dimensional aerofoils and three-dimensional wings; many other topics. 654pp. 5⅜ x 8½. 65465-6 Pa. $16.95

INTRODUCTION TO DIFFERENCE EQUATIONS, Samuel Goldberg. Exceptionally clear exposition of important discipline with applications to sociology, psychology, economics. Many illustrative examples; over 250 problems. 260pp. 5⅜ x 8½. 65084-7 Pa. $10.95

THREE PEARLS OF NUMBER THEORY, A. Y. Khinchin. Three compelling puzzles require proof of a basic law governing the world of numbers. Challenges concern van der Waerden's theorem, the Landau-Schnirelmann hypothesis and Mann's theorem, and a solution to Waring's problem. Solutions included. 64pp. 5⅜ x 8½. 40026-3 Pa. $4.95

LECTURES ON CLASSICAL DIFFERENTIAL GEOMETRY, Second Edition, Dirk J. Struik. Excellent brief introduction covers curves, theory of surfaces, fundamental equations, geometry on a surface, conformal mapping, other topics. Problems. 240pp. 5⅜ x 8½. 65609-8 Pa. $9.95

CATALOG OF DOVER BOOKS

CHALLENGING MATHEMATICAL PROBLEMS WITH ELEMENTARY SOLUTIONS, A.M. Yaglom and I.M. Yaglom. Over 170 challenging problems on probability theory, combinatorial analysis, points and lines, topology, convex polygons, many other topics. Solutions. Total of 445pp. 5⅜ x 8½. Two-vol. set.
Vol. I: 65536-9 Pa. $8.95
Vol. II: 65537-7 Pa. $7.95

FIFTY CHALLENGING PROBLEMS IN PROBABILITY WITH SOLUTIONS, Frederick Mosteller. Remarkable puzzlers, graded in difficulty, illustrate elementary and advanced aspects of probability. Detailed solutions. 88pp. 5⅜ x 8½.
65355-2 Pa. $4.95

EXPERIMENTS IN TOPOLOGY, Stephen Barr. Classic, lively explanation of one of the byways of mathematics. Klein bottles, Moebius strips, projective planes, map coloring, problem of the Koenigsberg bridges, much more, described with clarity and wit. 43 figures. 210pp. 5⅜ x 8½. 25933-1 Pa. $8.95

RELATIVITY IN ILLUSTRATIONS, Jacob T. Schwartz. Clear nontechnical treatment makes relativity more accessible than ever before. Over 60 drawings illustrate concepts more clearly than text alone. Only high school geometry needed. Bibliography. 128pp. 6⅛ x 9¼. 25965-X Pa. $7.95

AN INTRODUCTION TO ORDINARY DIFFERENTIAL EQUATIONS, Earl A. Coddington. A thorough and systematic first course in elementary differential equations for undergraduates in mathematics and science, with many exercises and problems (with answers). Index. 304pp. 5⅜ x 8½. 65942-9 Pa. $9.95

FOURIER SERIES AND ORTHOGONAL FUNCTIONS, Harry F. Davis. An incisive text combining theory and practical example to introduce Fourier series, orthogonal functions and applications of the Fourier method to boundary-value problems. 570 exercises. Answers and notes. 416pp. 5⅜ x 8½. 65973-9 Pa. $13.95

AN INTRODUCTION TO ALGEBRAIC STRUCTURES, Joseph Landin. Superb self-contained text covers "abstract algebra": sets and numbers, theory of groups, theory of rings, much more. Numerous well-chosen examples, exercises. 247pp. 5⅜ x 8½.
65940-2 Pa. $8.95

STARS AND RELATIVITY, Ya. B. Zel'dovich and I. D. Novikov. Vol. 1 of *Relativistic Astrophysics* by famed Russian scientists. General relativity, properties of matter under astrophysical conditions, stars and stellar systems. Deep physical insights, clear presentation. 1971 edition. References. 544pp. 5⅜ x 8½.
69424-0 Pa. $14.95

Prices subject to change without notice.

Available at your book dealer or write for free Mathematics and Science Catalog to Dept. GI, Dover Publications, Inc., 31 East 2nd St., Mineola, N.Y. 11501. Dover publishes more than 250 books each year on science, elementary and advanced mathematics, biology, music, art, literature, history, social sciences and other areas.